The Practice of Theoretical Curiosity

EXPLORATIONS OF EDUCATIONAL PURPOSE

Volume 20

Series Scope

In today's dominant modes of pedagogy, questions about issues of race, class, gender, sexuality, colonialism, religion, and other social dynamics are rarely asked. Questions about the social spaces where pedagogy takes place – in schools, media, and corporate think tanks – are not raised. And they need to be.

The *Explorations of Educational Purpose* book series can help establish a renewed interest in such questions and their centrality in the larger study of education and the preparation of teachers and other educational professionals. The editors of this series feel that education matters and that the world is in need of a rethinking of education and educational purpose.

Coming from a critical pedagogical orientation, *Explorations of Educational Purpose* aims to have the study of education transcend the trivialization that often degrades it. Rather than be content with the frivolous, scholarly lax forms of teacher education and weak teaching prevailing in the world today, we should work towards education that truly takes the unattained potential of human beings as its starting point. The series will present studies of all dimensions of education and offer alternatives. The ultimate aim of the series is to create new possibilities for people around the world who suffer under the current design of socio-political and educational institutions.

For further volumes:
http://www.springer.com/series/7472

Mark Zuss

The Practice of Theoretical Curiosity

Springer

Mark Zuss
City University of New York
Graduate Center
Seeley Street 250
11218 Brooklyn
NY, USA
zuss@earthlink.net

ISBN 978-94-007-2116-6 e-ISBN 978-94-007-2117-3
DOI 10.1007/978-94-007-2117-3
Springer Dordrecht Heidelberg London New York

Library of Congress Control Number: 2011934270

Printed on acid-free paper

Springer is part of Springer Science+Business Media (www.springer.com)

Preface

This project derives from the many sources of influence that have lured, diverted, and beckoned me. As an imprint of the passion and pursuit of knowledge, this text is only a trace of its abiding presence in my life, and, as I hope to illustrate, the social nature of our embodiment. Affirming a restlessness of mind and body, this project is one arbitrary point of convergence for many projects across the disciplines. It is intended as a recognition of the power of knowledge in the plural, a tribute and questioning of the places of questions in our contemporary world.

Among the impulses that motivate this writing has been the persistent urge to become intimate with libraries. Writing among others in reading rooms, especially in the magnificent main reading room of the New York Public Library, amid the shuffling of pages, has stirred a desire to wonder about the motivations and kinds of interests other than mine that are in progress. In the diversity of things knowable, my own inquisitiveness has frequently led me to pore over the volumes and materials left by other researchers at their tables. A sifting of documents and their watermarked weight of signs is always a project of images and text in encounter. Poring over older books or volumes little used summons for me a desire to decipher originary traces, our tribal language's filiations in sense and memory. Histories of time start up from margins, flee with associative thought, loosened from literal reference; from nerve roots still dark with place, under lamplight, the imprint of hand and instrument, the writer and reader sojourn under the auspices of this technology of hands, their internal gestures rendered in slow arcs, loops, and plumes of ink, type, and pixel.

It is not a history of writing or the writings of history that interest me here, nor how specific and always partial histories are inscribed by inflection, accents, and tropes in the trains of thought. Memories converge in the momentum of images, the writing of light and shadows in their constantly revisable readings and figurations. It is the lure of the unknown, the pulse of what remains unsaid and possibly unsayable. I remain a curious listener, perhaps even an eavesdropper too, to the murmur, the circadian, tidal impress on what is thought and written that stirs my own. It is the bodies' images seizing and plying folds of the visible and articulable; it is sometimes just the stumbling start of a rhythm, their distant but approaching sounds conjoined in a dissonance from which I hear strains of virtual songs.

I am curious about rain and prisms, the way the afternoon light outside my windows breaks through a wedge of the avenue's sycamore trees; about what is called chance and the play of the contingent in the mundane habits of my life; about the possible fortunes and futures of the human and its various hybrids; about ways of being and knowing that create what Foucault once suggested in thinking of a critical community; in technically mediated relations between the forms of life in their global commotion; in all practices and instances that question and resist the stamps and postmarks of an instrumental modernity; about new expressions of singularity and difference, corporeality, and becoming.

I wish to acknowledge the sources of support and encouragement for this project.

I want to thank my students and colleagues at the Graduate and University Center of the City of New York. Among them I wish to acknowledge the sustained support of Ken Tobin, as well as Gaoyin Qian, Lehman College, and Lakshmi Badlamudi, La Guardia Community College. The conversations, ideas, and enthusiasm of former and current students, including Christine Siry, Carolyne Ali-Khan, and Gene Fellner, have goaded me into thinking about the origins and logics" of inquiry. Most of all, to Lisa Lincoln, who has always encouraged this project and who accompanies me in crafting the future.

A "Just" Curious Introduction

Riding along in my automobile
My baby beside me at the wheel
I stole a kiss at the turn of a mile
my curiosity running wild
cruisin' and playin' the radio
with no particular place to go
Chuck Berry

It remains fashionable to be curious. In contemporary consumer and industrial countries, an alert, attentive mind is celebrated; in the modernist West it is acknowledged as a disposition receptive to novelty, knowledge, conversation, and image, responsive to all manner of stimuli in an exploratory, playful environment. Strongly associated within the affective and cognitive contexts of human development, creativity, and educational experiences, curiosity reigns as a particularly alluring marker of native intelligence. It is a sign of childhood's epistemic roots as they intertwine toward adult cognition and participation in liberal democratic citizenry. A blazon of modernity's own self-recognition, curiosity, signifying alacrity, a willingness to surprise, to welcome the unexpected, it represents, and is often uncritically applauded, as a universal facet of a thriving emotional and intellectual engagement with the world.

Infrequently questioned, this readiness to question the order and patterns of experience is rarely itself the subject of inquiry. As a peculiar turn of the mind, it pursues the singular, the odd, irregular, and novel aspects of the familiar, signaling an often discomfiting indeterminacy and lack of finality. The historical and cultural conditions that initiate, augment, and regulate it are left, curiously enough, largely ignored.

Lighting the spark that spurs research, curiosity is the spirit of an abiding inquisitiveness realized in distinct social practices. It is the intent of this study to attempt a historical profile of curiosity's places and appearances in contemporary life. Cultural dynamics, interests and relations in class-stratified and consumer-driven societies present distinct variations that, if not at odds with, are starkly differentiated from its older, theologically saturated, or emergently secular sanctioning and moral valences. The interests that motivate my own curiosity are evident in the selections and emphases made from the irreducibly vast resources and instances of its expressions. I advocate for an appreciation of practices that stir what I identify as a discerning critical curiosity. In each chapter I offer profiles of significant agents and practitioners of theoretical curiosity in order to illustrate how theoretical projects are always socially and institutionally conditioned. Moreover, my own quite interested curiosity delves the question of the generation of this elusive critical engagement and spirit of inquiry itself. I explore the elemental, perceptual, and sense qualities of experiential fields whose collective momentum can spawn sustained projects for the reinvention of everyday life.

In the questions it allows itself to ask itself the degree to which its passwords, shibboleths, and keywords condition the shaping of everyday life, theoretical curiosity offers up a portrait of the possibilities of thinking within a historical age. Since at least Hume (1965) it begins the process of articulating an existential disquiet. Theoretical curiosity contests cultural spaces for the persistence of skepticism, for an abiding and fundamental uncertainty regarding the nature and composition of our world of experience. Complementing and counteracting the force of doctrine, law, and habit, it generates an "uneasiness in its instability and inconstancy." For Hume, in his skeptical forays, curiosity "pleases because it produces belief, and removes uneasiness and doubt" (p. 655). As a quest, curiosity is paradoxical, at once an elusive and focused desire. Its cunning ruses and researches, though often oblique and indirect, work without preexisting maps, schema, or a priori category. It can be at work and play where something is hidden, anomalous, missing, or uncertain. Generating genuinely inquisitive questions and persisting in searching for their place among the phantoms and ephemeral streams of experience, practitioners of theoretical curiosity suspend expectation. It sets aside prior knowledge in systems of inquiry based on any certainty of the world and the order of things.

It is worthwhile to separate the forking paths of interest, wonder, and curiosity, of the forbidden from the merely marginal, and the objects of study one "dares to know" from those that pose too great a risk to the intellectual adventurer. This is a process of domestication of the unknown or unknowable, of what, Charles Nicholls (1997), in a biographical interpretation of Arthur Rimbaud's abandonment of European culture and literature for his African business exploits, regards as an "entropic dwindling of the unknown into the familiar" (p. 149). Many research programs and their research communities gain recognition and stature as they come into focus for their potential utility. These institutionally forced programs and methods of knowledge-work, as in genetics and nanotechnology, for instance, are instrumentally managed. All research converts "noise" into information in some manner, detecting pattern and symmetries. Researches in the technosciences today must ferret through vast quantities of "data" in the phenomena they study, pursuing the anomalous in the active cultivation of what appears as plausibly relevant and coherent within initial forms of randomness. At the historical beginning of the experimental method, Francis Bacon's (1996) inquiring mind could observe that "there is no excellent beauty that hath not some strangeness in the proportion" (p. 189).

Let me state here that while I distinguish theoretical curiosity and its forays from "ordinary" curiosity, this work stresses their common origin in our embodied perceptual orientation within situated contexts and "lifeworlds." I reject all dualisms in partitioning affect and reason, sense and intellect, and theory and practice. My inquiry here suggests that intellectual and theoretical curiosity inhabit no permanent epistemological or cognitive niches. They do not constitute or represent a unified "faculty" or essence of the mind. They partake simultaneously in processes engaging available historical networks and social resources insofar as they provide for or restrict a primary human desire for knowledge. Sustained, purposeful, or apparently useless "blue sky" inquiries are gateways to what, for the empiricists and skeptics alike, was the fancy or imaginative faculty.

One intention of this project is to leave open the question of the effective purpose and social influences that support, sponsor, and control both purportedly speculative and utilitarian pursuits of knowledge. I adamantly wish to avoid contributing in any way to curiosity's domestication. This inquiry into inquiry attempts another track, tracing movements of thought, and, tempts courting a paradox: accepting a willful purposelessness and holding back of any necessary utility, value, or use of the inclination for theory, including in its most abstract and speculative practices, as for instance in mathematical or cosmological quests, as they hew their own specialized, circuitous paths. This jettisons the millennial shadows cast over the quality of this uniquely social and individual experience in all its morally charged attributions and aspersions. The perennial trespasses and explorations associated with it have borne the brunt of divergent forms of disquiet, among them the epistemological and political issue of what comprises useful knowledge. In attempting to question the will to question itself, I wish to hold up for critical scrutiny certain philosophical and political resonances of its polyvalent historical reception, in its tribulations, celebrations, and condemnations alike.

In attempting to question the will to question itself, I hold up for critical scrutiny its historical and contemporary resonances. I wish to demystify curiosity's peculiar status in contemporary culture in general, and its intellectual place, especially in its role in questioning the nature of everyday life and practice. I adapt its historical and phenomenological nuances, to advance a socially enabled *critical curiosity*. This is intended as a way of continually attempting to make sense of the unchartable continua of experiences, perceptions, paths of inquiry, and thought experiments within specific communities of contemporary cultural practice. I offer profiles of the ways in which some distinctive contemporary cultural and technical practices, in particular pedagogy and technics, including genomics, artificial life, and exobiology, come to public attention. I want to place our theoretical adventures in question, placing accent on the nature of everyday life, its rhythms, order, and potential for transformation.

Curiosity deflects its own historical interests. In contrast to all accounts that erase curiosity's historical imprints, and those that reduce it to a merely developmental feature exhibiting distracted attention, I claim that it functions under suspicion. As an alluring quality of contemporary cosmopolitan personae, as is recognized in an open-ended spirit of discovery, and of childhood's sensory immediacy, I wish to marshal evidence that the force of theoretical curiosity remains as vitally interested, implicated, and engaged in contemporary practices, particularly in the technosciences, as it was in the lives and times of such figures as Bacon or Maupertuis, Voltaire or Humboldt. I believe that curiosity is a genuine affective and cognitive resource for communities of practice; it is not a neutral or naturalized aspect of a psychic interior or human ontogeny alone. As a socially sanctioned practice, granted varying and often inconsistent and contradictory degrees of legitimacy, it remains fraught with ideological, ethical, and political charges and significations. It is not a permanent, ahistorical process, psychological state or trait in the everyday construction of objects for the knowledge of individual practitioners. I think it is useful to adopt Paul Rabinow's commitment to questioning the formation and tendencies implicit in its contemporary manifestations.

As an enduring, if mutable, passion of the mind I chart some of its filiations and alliances with other formative cultural practices, including economic, aesthetic, biological, pedagogic, and philosophical activities and traditions that have established themselves as the panorama and horizon into which contemporary culture appears to contemplate and realize its futurity. Favoring a critical imagination and elemental curiosity in the wake of a general cultural tempering or "taming" of this singularly resilient and elusive passion, I attempt to provide an appreciation of curiosity's presences and reception in specific domains of thought and activity as a sort of tutelary deity or demon across all disciplinary endeavors.

In my pursuit of the new frontiers of inquiry, I try to demonstrate how research and design initiatives increasingly merge market and academic alliances. Academic capital and technoscientific projects like artificial intelligence's "nanobots" or patented genes, instrumentalize theoretical inquiry. Modes of investigation that are most likely to find commercial application are selected for funding, prestige, and promotion. The turn toward the pharmacological and genomic "reading" of the human genome, unprecedented realizations of "biocapital" and digital power, as well as the peculiarly insistent, restive searches for other forms of life, whether earthly or extraterrestrial, are distinctive instances in an increasingly ubiquitous contemporary process of hired hands and putting theory up for sale. It is a historical moment marked by a ravening instrumental imprint on what is at once knowable, how it becomes known, and what remain unknown and unquestioned.

In its puzzling of problems and its subversion of conventional wisdom and informed fact, theoretical curiosity's interdisciplinary methods are often eclectic. It is also a restless force within disciplinary knowledges. Theoretical curiosity is one of the conditions of possibility for the growth and change of human knowledge within and across fields. Never content with the status quo of a contented, well-fed repertoire, it is at minimum a latent social impulse resistant to remaining within existing human limits and symmetries imposed upon an unknowable, wholly astonishing infinitude. It is a pedagogical encounter with an unfinished, open-ended totality of the possible.

Despite its cultivated abundance within intellectual labor, the theoretical and cognitive drive should not be idealized. It is certainly not the determining factory or "destiny" of historical knowledge, as Blumenberg (1985) states, despite its singular part in preparing the ground for the ascendance of modern technological rationality. It is certainly in acute tension, however, with finalisms and all forms of intellectual, epistemic, and political closure, as I hope to make evident in discussing the contemporary cases of genetic and informatic paradigms in Chapters 6 and 7, in the singular case of the intellectual ferment in the micropolitics and theoretical practices preceding and surrounding the Events of 1968 in France in Chapter 4, and in the formulation of alternate pedagogies of concepts, in educational practice and in philosophical critique in Chapter 3. Final causes, purposes, and teleological projects are at odds with a theoretical curiosity that suspends business as usual, questioning the operating principles and practices that motivate them. It works against the grain of any will to totality. Digging into the very ground that supports them, practitioners of a critical and partisan theoretical curiosity take the risk of falling into abysses of uncertainty of their own making and discovery.

Theoretical curiosity is an enduring aspect of Enlightenment values that refuses the quietism of the contemplative, timelessly speculative life. It is also, pace Gramsci, necessary for this practice to continually question its own foundations and limits. Critical projects conducted in this vein are to be found intervening into the givenness of systems, laws and paradigmatic operations that generate only abstracting structures and forms without regard for embodiment. The critical pedagogy of curiosity that this project sponsors resists any resurgence of a disembodied, purportedly objectivist and non-partisan practice in speculative thought, in theoreticist formalisms and positivist claims for disclosing the nature of "real," of all kinds, as abstracted from material historical practice and experience In thinking through how a great diversity of inquiries traverse shared social experience, I suggest that it partakes of phenomenological qualities, insinuating itself in, and often against, the grain of conceptual order, category, and disciplinary boundary. I believe that as a quality of intellectual practice, it names germinal processes at work in the formation of the sensible, of the basic givenness of the natural and elemental worlds of perception and experience. It is operative in the presentations of the qualitatively sensible depths of experience, in how sense itself becomes present, attentive, careful in its probing pursuits. Borrowing from the rich insights of the phenomenological tradition, especially the researches into perception by Merleau-Ponty in Chapter 5, I open up old and persistent questions. This project is motivated and encourages a return to the beginnings of knowledge in perceptual "wonder" first remarked upon by Aristotle and evident in the Presocratics's protean inquiries. If our long range theoretical, scientific and professional pursuits are generated by more than the perpetuation of existing practices, it is necessary to open the shades of objectivity to allow for what Merleau-Ponty (2002) artfully describes as a "primordial layer at which both things and ideas come into being" (p. 255).

My first three chapters are intended to portray signal developments and historical conditions in which theory and curiosity are both at play and in contest. Each chapter of this first section and the text as a whole is self-contained to allow for being read independently and read out of sequence. In Chapter 1, "First Questions," I provide a genealogical silhouette of the shifts in value and significance of theory and its relation to curiosity within some of the dominant cultural dynamics in the ancient Presocratic, classical Greek and medieval eras. Theoretical inquiries have both served and severed knowledge gained from observers' perceptual sense. Viewed from a cross-historical perspective theoretical knowledge has mediated the aesthetic and practical working knowledges, whether, for instance, in crafting, artisanal, medical, and hunting and fishing practices; intellectual, symbolic, and abstract knowledges and the value of "pure" theory, unencumbered by perception and sense and curiosity, have ambiguously presided over the working knowledge in embodied collective experience and cultural memory.

In tracing this genealogy I was repeatedly surprised to find a lasting and deeply ingrained identification of curiosity, theory, and sight. Theoretical curiosity participates in what I prefer to call a play of light across the centuries and epistemic regimes of thought. It is a prism, mediating the empirically visible and objectified world of the natural sciences and the unexamined life of the everyday. Since the classical era and through both medieval and early modernity, as well as up into what might now constitute our technoscientific postmodernity, the visual has remained the dominant mode and figure for the acquisition and apprehension of knowledge. Vision and the optical tools, maps, charts, lenses, and scanners, for instance, have practically and figuratively focused the attention that curiosity is claimed to disperse.

In the first chapter I detail aspects of classical pedagogical "schools" in their pursuits of what I identify as one of the forms for the practice of theoretical curiosity. I emphasize the influence of the Miletian Presocratics in the reception, transmission, and care for "first questions," the realm of inquiry into causes and processes of all things in nature. I consider aspects of ancient doctors' capacious inquiries in the context of a considerable and enduring ambiguity regarding the senses and intellectual knowledge. From Thales and Anaximander into the classical era, there is a prevailing identification with an essentially speculative theoretical curiosity. Their mainly specular interests in cosmological and natural processes, while standing outside of time and place, however, also appear to have often remained grounded on earthly empirical observation and pursuits, including the building of dams, and premises intervening into such diverse fields as medicine, military strategy, and vital sea trade routes. Visibility persists as a point of theoretical interest in the succeeding Academies and Lyceum, exhibiting what can be thought of as a more theoreticist and disembodied turn that is most readily found in the Plato's realm of intelligible forms or Aristotle's Metaphysics. A general "downgrading" of perception accompanies the resurgence of the transcendental idealism of the era. It is a robust rationalism that will shed all empirical clothing to bathe in what, later, Descartes would praise as the "natural light" of intellectual intuition grounding true knowledge of the real. This chapter adds a brief review of some of the Latin and medieval attitudes toward curiosity. From the early church fathers from Augustine up until Nicholas of Cusa, I describe the practice of disciplining and almost universal condemnation of curiosity as an errant straying of the mind from devotion to the true Christian faith. Across distinct eras, the spirit and practice of inquiry has been served notice to appear at what Hans Blumenberg (1985) called the "trials" of theoretical curiosity. It has been subject to changing historical regimes of thought and disciplinary practices; it has been valorized, accepted, and condemned alike.

My second chapter, "Taming of the Passions," portrays how curiosity is newly respected and enlisted within colonial enterprises during the Age of Discovery. Curiosity is acquitted; no longer luring straying minds into deadly sin, it becomes incorporated into nation-state formation and early industrial and technical enterprises. As in Hobbes (1968), it is a "singular passion" (1.6, p. 124) within an ethos in which accumulation, acquisition, and even avarice become laudable. Cabinets of curiosity and the collectible artifacts of colonial plunder are iconic of a curiosity also rendered in useless exotic entities rather than activities of the mind and senses. In this chapter I also emphasize the emergent Renaissance commitment to bringing form out of the potential of material substances through the exertion of a theoretical and intellectual vision. In the time of Leonardo, the creative act of thinking with and through substances was displayed in melding technical knowledge with mathematical and geometric theory. Mathematical models gain stature as models of true knowledge in the emerging sciences after Galileo. Their role in abstracting value from the material world of labor is considered in terms of the rise of mercantile capitalism and its dominant modes of forming subjectivity. During the same era, however, it is also an era witnessing a rare merging of the rational and empirical, the intellectual and the artisanal. This chapter also argues for a consideration of curiosity as a movement of embodied thinking or thinking bodies, adapting Spinoza's tenets regarding the affects and passions. I complete this review by considering the Encyclopedists, whose systematic, methodical practice displayed ambivalent and contradictory attitudes toward the place of theoretical curiosity in the Enlightenment's projects.

Chapter 3, "Pedagogies of Curiosity," considers the pertinence of a critical curiosity to questions of pedagogy and socialization in modern life. It examines psychological reductions of curiosity as accomplices in a social and behavioral science tradition that has valorized its epistemological inventiveness over its "dispersed" or "ingenious" instances and practices. I survey the cognitive and experimental models attempting to establish normative and quantitative parameters for curiosity. Psychological premises, including "drive" theories of curiosity, are challenged as attempts to commodify the desire to know and possibly change a problematic world into an ahistorical, natural essence of a purported human nature. The enduring ambivalence and conflation of sensory and epistemic curiosity is examined as reflected in theories of desire. Freudian premises regarding the nature of curiosity and desire, as exemplified in his study of Leonardo, are critiqued from the vantage point of pragmatist thought and contemporary materialist feminist articulations. It surveys Freirean critical pedagogical approaches and their invocation of a sustained epistemological curiosity. The shifting values associated with curiosity are scrutinized as an aspect of its institutionalization in public education that render it hostage to an actuarial surveillance of the working class. With Freire, I affirm theoretical curiosity and engaged inquiry as primary elements in any genuine pedagogy of freedom.

Chapter 4, "The Sphinx," pursues the question of everyday life as a frontier of curiosity. It begins a focus on theoretical practices *within* specific modern and postmodern practices and problematics. Here I consider the everyday of modernity as the untheorized and unquestioned domain of experience and the problems it brings to bear in its repetitions, patterns, and temporality. The foci for this question are the theorists of everyday life, in particular Henri Lefebvre, Heidegger, Althusser, and the Situationists. In a twentieth-century philosophical exploration of everyday life, Henri Lefebvre (1991) writes that critical reflection and inquiry "implies an attempt to interfere with existing conditions and an awareness of other policies than those in force" (p. 174). It is the spy of culture, questioning the possibilities of how the order of things might be differently imagined and configured, suspending the finality of all answers theoretical, pragmatic, and empirical. Curiosity often makes its appearance unexpectedly; it demands recognition as a potentially disruptive desire that troubles the sleep of tradition and privilege.

This chapter reviews the innovative departures from orthodox historical materialism in these theorists and their particular ways of accentuating the elements of contingency and the aleatory in social structures and relations. As projects of theoretical curiosity, these formulations present utopian moments, ruptures, irreducibles, and remainders in everyday, quotidian experience. This chapter advocates a critical curiosity as a sustained uncertainty and perpetual inquiry into the transformative potentials and virtualities present in daily life.

Chapter 5, "Curiosity and the Question," is an appreciation of curiosity from a phenomenological perspective. It is the most "theoretical" section of the book in that it speculates on the enduring questions of the relation between perceptual sense and the growth of knowledge raised in earlier chapters. I develop a critical approach to the descriptively rich phenomenological method and tradition. This chapter engages with Merleau-Ponty's critical interventions into contemporary thought in interpreting theoretical curiosity as an "expression" of the primacy of perception and sense as the ground for all intellectual, linguistic, and conceptual pursuits. Sense knowledge and perception are the fields, source, and horizons of the "natural attitude" of everyday experience. They are also the points of return for our collective and embodied lives.

A critical appreciation of the phenomenal qualities that generate theoretical practice allows for surpassing all dualistic and hierarchical models.

Perception, concepts, and the "cognitive drive" that often delimits theoretical practice are rewoven into a continuum in which they form intersecting layers or folds in the textures and texts of our becoming. I follow through with a provisional critique that advances a "new materialism" taking flight with premises of Merleau-Ponty and especially Gilles Deleuze's germinal thinking about the production of concepts. I present a reinterpretation of the grounds of knowledge within the horizons opened by our perceptual embodiment vitalized with a renewed regard for what might constitute a critical ontology. From Deleuze's proposed "planes" of immanent thought, an infinitely productive ontology precedes all epistemological and representational systems or models that derive from it. In my raveling together concepts from phenomenology and post-structural anti-humanism, this chapter offers a glimpse of a future freeing of practices from the cul-de-sac of rationalist and empirical dualisms, epistemology, and representational capture.

Chapter 6, "Thinking Life," applies tenets of a critical curiosity to new techno-scientific practices. It questions the limits of a theoretical curiosity of genetic technologies that challenges the divides between human and non-human life-forms, as well as making increasingly indeterminate the boundaries between forms of life and the inanimate. It provides a profile of the technical apparatus now being developed that extends the power of the human and the de-differentiation of species-being in the genomic paradigm. It contextualizes the genetic revolution within an appreciation of the salience of a perennial technics within human cultural organization and practice. This chapter claims that the new genomic models compel a shift away from public sector and democratic polities in favor of a technical sovereignty over life. It promotes a new critical, "zoographic" practice that addresses the extensions and changes in the relation of the body to the world and the nature of experience, reviewing germ-line and somatic cell developments in genetics and reproductive technologies affording class-marked bodies, health, and longevity.

Extending the arguments on curiosity, technics, and zoographic inquiry in Chapter 6, Chapter 7, "Minds, Limits, and Spaces," critiques recent investigations in artificial life, exobiology, and "emergence" in bioinformatics. It reviews the intensified politicization of scientific research in the United States, specifically of NASA, and contextualizes the exobiological search for forms of life on Mars and elsewhere within developments of the technosciences and global capital markets. In the contemporary explorations highlighted in these two chapters, original thinkers today can be found delving into the realm of the infinitely small, in particle and quantum physics, as in the work of Gell-Mann, Feynman, Kauffman, and others, as well as in the infinitely large, in the search for origins, patterns, and processes that have generated and sustained cosmological systems. My emphatic concern here remains the situated and embodied activity of the mind in social relation, as communal practice and interest, generated by and generative of sustained social relations, including research programs and paradigms. I consider these multifaceted facets of curiosity as socially and historically enabled artifacts, representational imprints of contemporary power and desire. Drawing from the phenomenological premises of Chapter 5, I offer a critical interpretation of the question of the limits of epistemic curiosity, leaving open core question of what constitutes forms of life, the borders of the animate, inanimate, and lifelike, emergent systems. I conclude with thinking our experiential beginnings, of who and how these novel fields of theoretical inquiry are giving shape to the future.

Contents

Chapter 1
First Questions

The historical fortunes of speculative thought and theoretical curiosity begin in laughter. In Plato's *Theaetetus*, a discussion of the place of the philosopher becomes the subject of debate between Theodorus and Socrates. The philosopher's persona is described as an absent figure in the marketplace, the polis, and the affairs of everyday life: "the philosopher grows up without knowing the way to the marketplace, or the whereabouts of the law courts or the council chambers or any other place of public assembly." The philosopher is unaware of laws, written or oral, as these are things "he never sees or hears." A drifting, ethereal existence, this benighted existence permits him to exist without knowledge of "questions of birth," genetic inheritance, or class status. He is so utterly removed from the concerns and cares of the quotidian that

> he knows not even that he knows not; for he does not hold himself aloof from it in order to get a reputation, but because it is in reality only his body that lives and sleeps in the city. His mind, having come to the conclusion that all these things are of little or no account, spurns them and pursues its winged way, as Pindar says, throughout the universe (Plato, 1997b, pp. 192–193 [173e]).

Socrates describes the prototypical quests of the Ionian natural scientists. His description is apt for the entire tradition beginning in the sixth century with Thales and Anaximander: Pursuing "the deeps below the earth and in the heights above the heaven; geometrizing upon the earth, measuring its surfaces, astronomizing in the heavens; tracking down by every path the entire nature of each whole among the things that are, and never condescending to what lies near at hand" p. 193).

It is Thales who serves as the target of ridicule for Socrates's famous anecdote. It is a tale that accents the Occidental, classical heritage, including the many schools of skepticism, in recounting how laughter greeted the theoretical turn of the mind. He retells a legend:

> They say Thales was studying the stars, Theodorus, and gazing aloft, when he fell into a well; and a witty and amusing Thracian servant-girl made fun of him because, she said, he was wild to know about what was up in the sky but failed to see what was in front of him and under his feet. The same joke applies to all who spend their lives in philosophy (p. 193 [174b]).

M. Zuss, *The Practice of Theoretical Curiosity*, Explorations of Educational Purpose 20, DOI 10.1007/978-94-007-2117-3_1, © Springer Science+Business Media B.V. 2012

A comic presence in the public mind, the philosopher is an object of derision. Oblivious to the everyday, to his neighbors and to the busy-body concerns of the world, its gossip and its vanities, the philosopher has chosen, for Socrates, to pursue the nature of virtue in a world permanently inhabited by evil. For Socrates, the "question he asks is, What is Man? What actions and passions properly belong to human nature and distinguish it from all other beings? This is what he wants to know and concerns himself to investigate" (Plato, p. 193 [174b]).

The desire for knowledge is an enigma of history. Seeking understanding of the elements, processes, and structures of the inanimate and animate dimensions of our world, researchers have continually confronted knowledge's human limits. Sustained forms of inquiry, the pursuit of the practices of theoretical curiosity that I wish to profile here, situate the convergence of contingent limits; in challenging or defying them, certain research projects, however local their scope and methods may be for their counterparts in contemporary inquiries, resist the finitude of human knowledge. Researches of this kind, from Odysseus's voyages beyond the then-known navigable waters to the astrobiological exploration of unique life-forms, mark points of possible discovery, and invention or intervention. Projects of theoretical curiosity manifest a resilient and desiring knowledge, the will to surpass existing domains of knowledge, and the quality of our embodied relation to the world. I begin with a review of some of antiquity's embrace of an immense diversity of research and pursuit of foundational principle, of first questions, of nature and life.

Theoretical curiosity conventionally connotes inquiries that test and resist imposed limits to thought within existing forms of practice, method, and intelligibility. Hans Blumenberg's (1985) sweeping historical critique of the "trials" of theoretical curiosity documents its perduring contestation within and across shifting sociohistorical milieu. I contend in the following historical profiles that a resistance to the containment and limits of knowledge continues to be met into contemporary inquiry. The tension between speculation and utility, evident in the context of the concerns of the ancient "doctors," pertains to issues in the pursuits of theoretical curiosity in the technosciences in late capitalism.

From classical *theoria*, an originary, Aristotelian wonder, through the trial of Socrates, its condemnation by St. Augustine and the early church fathers, onto the confrontations brought to bear in the wake of the challenges presented by Galileo, Copernicus, and Bruno, theoretical curiosity has borne epistemic, moral weights and balances. In the unprecedented ventures and discourses of early modernism, discussed here and in the following chapter, a contretemps was maintained between an emergent empirical practice, *scientia,* that would valorize it, and the prevailing theocratic and patristic order that confined and regulated it through its canons, inquisitions, and curricula.

In the following, necessarily brief summary of the place of theoretical curiosity in the classical context, I will address only two major strands of debate and interest: first, the relation of the senses to the intellectual functions, the sensible to the "intelligible" or noetic and timeless realms of ideas, and second, the question of the limits and purposes of the diverse and insistent quests of human knowledge. Starting with the formidable range, depth, and variety of some of the important "schools" of thought in antiquity I consider these primary questions and their influence throughout the middle ages as well as upon our own possible reflection on presiding modernist practices. This will require description of the conditions and contexts that promoted, shaped, or constrained research and discovery. Speculative researches usually denied and continue to disparage the empirical that are their only possible actual foundation. "Pure" speculative theory, as in forms of ancient and modern cosmology, exists outside of any but an ideal relation to the historical desires, needs, and aspirations of its age. Speculation might need to exist as an intuitive moment within the dialectics of discovery and invention, but cannot genuinely develop without reflecting, at least implicitly, a desiring knowledge. Without falling into dogma, blind faith, or chicanery, without the desiring for knowledge of a body, a body whose senses have been shaped by its inheritances, I believe that no sustained inquiry can ever be maintained.

Curiosity in Classical Inquiry

In the classical context Cicero interprets Ulysses's journey past the Siren's island as a struggle to resist the lure of knowledge, claiming that

> it was not the sweetness of their voices nor the novelty and diversity of their songs, but their professions of knowledge that used to attract the passing voyagers; it was the passion for learning that kept men rooted to the Sirens's rocky shores (Cicero, 1994, pp. 449–450).

In his protean wandering Odysseus personified a confrontation with the limits of knowledge of the world and a metaphor for the moral trials of self-discovery. From the classical era through to Dante and the early Renaissance, as discussed in the next chapter, the continually interrupted return of the hero and his crew occupied a unique place in Western epistemological figuration.

As "allurements of knowledge," the Sirens's temptations are irresistible. Drawing its linkage to childhood and to insatiability, this passion stalked all human endeavors. For classical thinkers, curiosity, as *theoria,* a passion for knowledge, would, as an ideal, occupy an autonomous realm. Cicero's comments direct attention to these facets claiming

> so great is our innate love of learning and of knowledge, that no one can doubt that man's nature is strongly attracted to these things even without the lure of any profit. Do we notice how children cannot be deterred even by punishment from studying and inquiring into the world around them? (Cicero, 1994, p. 449).

At the same time, Cicero admonishes "a passion for miscellaneous omniscience stamps a man a mere dilettante" (p. 451). He gives voice to a commonly held belief of his era that a passion for theoretical curiosity must correspond and relate to the conduct of individual life's obligations and responsibilities, in a balancing of the forces of "necessity and freedom, of public duty and private interest" (p. 452). Cicero claimed that a knowledge of nature, *physis*, guarantees a liberation from fear. In particular, the paradigmatic and frequently contentious study of astronomy permits humanity to attain a degree of autonomy and agency. It grants a measure to humanity's place, allowing it to participate rationally and in balance within the spheres of social justice and the order of the city. Hans Blumenberg regards Cicero's restrictive inquiry as a facet of an abiding "economy of the active subject" in which the "appetite for knowledge of hidden and marvelous things" in which humanity might pursue its essence is permissible and appropriate "only in the leisure that is appropriate vis a vis other human and civic demands" (Blumenberg, 1985, pp. 280–81).

In classical philosophical thought, inquiry bridged the sensible world with immaterial forms. It presided over speculation and empirical explorations of nature, matter, motion, and change. Theoretical investigations, from the Presocratics on, traversed the limits and linkages between visibility and invisibility, sense experience and unchanging being, to disclose unities, symmetries, and harmonies. Jacques Brunschwig (2000) asserts that these "natural doctors" "plunged headlong into an assault on knowledge of the whole, of the principles on which everything is based" (p. 29). An economy of attention is apparent in the multiple tributaries of classical thought. As a guiding premise, this concern and caring for the direction and expenditure of energy, thought, and emotion consolidates otherwise-disparate communities, or schools of thought. It animates researches from Cicero and Plotinus on to Augustine. Skeptics, Stoics, and Academics shared variations of a common interest in directing the gaze of theory toward specific objects of contemplation and practice. For Plotinus, this purposiveness of mind formed a central tenet. In his *Enneads* it constituted what Pierre Hadot (1993) identifies as a "conversion of attention" through continual contemplation, of the unity and simultaneous coexistence of multiple and intersecting levels of being, knowledge, intellect, and experience. In Plotinus, as in Academic Skepticism that followed after Plato's death, this caring for attention, as a disciplined *askesis*, was directed and structured as a "turning toward transcendence" (p. 5). In Blumenberg's account, theoretical activity within the classical episteme was constellated in an economy of thought encompassing the "breadth and variation of Hellenistic formulas" in their passionate confrontation with the "threat posed to human *eudomonia* by the appetite for knowledge" (p. 279).

Distinctions and disagreements about the place and purpose of theoretical practices, however, abound classical discourse. Aristotle (1998) provided the well-known claim that "by nature, all men long to know," and that it is "because of wonder that men both now and originally began to philosophize." In the context of the *Metaphysics*, he advocated for a purely theoretical attitude as a disinterested end in itself. He naturalizes the desire to know, arising spontaneously from of an originary wonder. Finding themselves in ignorance, humanity "pursues science for the sake of knowledge and not for any utility" (Alpha 1, p. 980a; Alpha 2, p. 982b.). Aristotle is perhaps disingenuous here, acting in the interest of pursuing his own new course of natural inquiry, reinterpreting Plato and the earlier, less systematic observations and analyses of *physis* and the grounds for knowledge by the Miletian "natural scientists and physicians."

G. E. R. Lloyd (1970) provides a rich inventory of the varied projects and themes pursued in antiquity. He finds no universally accepted view of the nature of scientific inquiry. The criteria for research and inquiry varied considerably in what could constitute a branch of knowledge as opposed to arbitrary speculation. Important differences did exist between the decentralized cultures of the Presocratics and

Athenian schools. Plato and Aristotle disagreed, significantly, on the value of observation and both argued against the atomists and Empedocles over final causes. For Lloyd, "we may conclude that curiosity and the desire for knowledge, both theoretical and practical, were the most general motive forces behind the investigations of nature in antiquity" (p. 138). This core issue, of utilitarian as opposed to purely speculative or contemplative knowledge, as Aristotle provides in the *Metaphysics*, kept open the kinds of inquiries and the degree to which empirical or experimental methods were employed. In Lloyd's extensive review, "the idea of undertaking deliberate research was a natural extension of the curiosity that motivated all Greek scientists to a greater or less extent, but first the Hippocratics, and then Aristotle especially, showed what it was to set about collecting detailed information in a particular field of inquiry" (p. 145).

A life of contemplation was the supreme life. Both Plato and Aristotle affirmed the absolute value of abstract and formal principles. It is what Lloyd refers to as a commitment to a transcendent realm of forms and ideals, a "positive glory in the ideal of the pursuit of knowledge for its own sake" (p. 132). Their work was radiated by an immaterial sun. They would reside apart from human affairs in contemplative lives in which reason cultivates true excellence. Contemplation, speculative thought, and study of nature, which included the beauty and order of universe shapes an orderly and noble character. An ambiguity arose in applying this ideal life to inquiries of nature. It was resolved to Aristotle's satisfaction by asserting that the intelligibility of an ordered, finite, and eternal universe could be discerned in the heart of the temporal and changing world of nature. Through the careful taxonomic recording of animals and plants, as aspects of the sensible world, knowledge of the principles, forms, and structures of living things would be attained and conform to the ideals of a life of speculative theoretical curiosity.

Theory in the Presocratic, Ionian, and later Academic philosophical milieu, or "schools," incorporated dimensions of human experience that do not generally pertain to it in contemporary thought. Theory originally connoted merely a way of seeing. According to Bruno Snell (1953), *theoros* meant a participation in an event as a spectator, as one who would watch the Olympiads. Its later derivations led to associations with contemplation and the acceleration or intensification of the usual functions and purposes of the eyes. It stressed the workings of the ocular in itself, in its apprehension of objects, distinguishing it from cognates that played down its specific role of taking in and appropriating a visual image.[1] Specific and presiding historical "regimes" will establish closely entwined relations between the sense of sight, curiosity and theoretical practice.

A theoretical attitude did not displace the significance of artisanal and manual labor. It also did not mark a boundary or hierarchy between affect and cognition. Snell interprets the shifting valences of *noos*, to realize, or to see in its true colors. In the modernist epistemological interpretation, as in Kant's transcendental schemas, *noos*, the aspect of universal mind, is the recipient of images, judgment, and the formation of conceptual knowledge, presiding over intellectual work. It is also a site for emotion and affective identification. *Noos* overlaps with *thymos,* the source or "mental organ" causing the reception of emotion, evading a dualism and cognitive divide later imposed upon theoretical speculation and inquiry and sensual, material experience (p. 14). Noos was an important concept in the classical world, indicating a seeing with the mental act that accompanies it.[2]

For Hans Gadamer (2001), following Aristotle, these keywords indicated an appreciation of the indissoluble link between thinking and sensing. In Gadamer's review, classical practice did not set up an opposition between the "relativity of sense perception and the absoluteness of thinking" (p. 106).[3] The place of sense perception in relation to the supersensible, the invisible, or the intelligible has vacillated within a philosophical spectrum arising in the classical context. For Aristotle, the realm of intellectual activity is also the domain of touch and the immediacy of perception. Aristotle's corpus exhibited an unwavering theoretical interest in the generation of living things. It is a line of inquiry I give special attention to in Chapters 6 and 7 in the context of genetics, artificial life, and exobiological research in contemporary technoculture. Over the immense range of topics considered, Aristotle's regard for the relation of the soul and the body shifted. By the time of the *De Anima*, his Platonism allowed for an appreciation of how the soul is expressed through bodily and sensory means, as in the heart. The soul and the body are no longer absolutely opposed as different substances. Form and matter comprise combined or common substances. In terms of the intellect (*noos*), where the essences are known and seen, everything, including the realm of the senses, becomes a possible object of thought.

Aristotle finds a correspondence between the five senses and a proper object, a particular medium or element of nature and specific organ. Sensations are always conveyed through the formation of a mean between extremes of magnitude, as in darkness and light, the raucous and the quiet, the bitter and sweet. As Daniel Heller-Roazen (2007) explains, "the sensible object, its medium, and the organ of its apprehension thus constitute formally distinct elements of sensation, which combine, in the event of actual aesthesis, to form a single experience" (p. 26). He discerns a puzzle in the order and hierarchy of the senses as presented in this seminal text. Last in his description of the senses, touch, for Aristotle, is the generative root of our being and knowing the world. In the *De Anima,* Aristotle (1966) states that touch is "the primary sensation that belongs to all animals" (2.2.413b4–5, p. 557). Roazen raises a question pertinent to the phenomenological accent I wish to place on any genuinely theoretical enterprise in its immersion or relation to forms of sight and sense. For Heller-Roazen

> sensible life comes into being with the presence of the power of touch and inevitably end, with perfect symmetry, in its absence. A difference of hierarchy thus separates the tactile faculty from the other senses: the existence of one founds that of the remaining four. But touch seems distinguished still further, on grounds that trouble the sense in which it can be called, in the terms of the De Anima, a sense at all (p. 27).

The subordination of the sensible to the intelligible, of appearance to timeless forms, from the Platonic dialogues and through Christian patristic doctrine into modernity, as discussed in later chapters, also entails the significant defamation of lived, sensuous experience. In the classical context of Plato true knowledge is a pure seeing of the non-sensible. From a phenomenological perspective, as John Sallis (2000) observes, this primary fracturing of experience and knowledge reflects an opposition between a epistemic, intellectual, or noetic understanding and an aesthetic regard, the "apprehension of things (or of the condition of one's own body) by sight, hearing, taste, smell, or touch" (p. 31). This has formed a deep and abiding historical and cultural chasm between epistemic and aesthetic approaches to inquiry. Noetic knowledge ascended as an intuition of the truth. Vision has held special status in a hierarchy of forms and qualities of knowledge. As a pure seeing, *eideani* indicated knowledge of essences. An intellectual vision, an *intuitus mentis*, will return in modernity with the potent disregard for sense knowledge as deducted by Descartes and the currents of rationalism of his era. I will return to historical uses or what Jonathan Crary (1999) describes as "regimes" of the visual and to theory as a form of sight in my consideration of technology and genetic techniques and in consideration of potentials for social change in "societies of the spectacle."

Access to enduring knowledge, the intelligible, has been questioned and challenged throughout the arguments of the philosophers since antiquity. One orientation allows for a mediation or trafficking between the sensible and the intelligible, between manifestations and differentials in actuality and potentiality, as in Aristotle. Since Plato, a metaphysical tradition has posited a limitless realm of the noetic or intelligible, an immaterial supersensible dimension that is inaccessible to embodied, human knowledge in itself. It is unchanging and timeless truth of pure forms, an ideality which "cannot be seen, heard, or otherwise sensed. And yet, it can be such as in some manner to accompany things seen or heard: as when, reading or hearing a word, one apprehends its sense" (Sallis, 2000, p. 31). Sense challenges this fundamental opposition, in being unlimited itself.

This sense of sense itself is a reevaluation of our core values and concerns about knowledge and experience. Placing perception at the leading edge of the thirst for knowledge and inquiry resituates an entire epistemological tradition. It brings to trial the legacy of a rationalism that denies the embodied mind. The fracture and division, often noted in many histories of thought, leaves the tracks of a philosophical and cultural mode of not-knowing, one that was, as Renaud Barabas (2006) writes

> constituted on the basis of a fundamental experience or an astonishment that leaves no room for the specificity of perceptual experience and in which it was ultimately concealed. The task of a philosophy of perception is therefore not to attempt to appropriate perception by means of categories at its disposal but to allow itself to be reformed through contact with perception; it must not conceive of perception so much as to conceive according to it (p. 18).

In the admittedly vast scope of inquiries ranging from the Homeric era to the Academics of classical thought and practice, the Presocratics mark a starting point for consideration of the fortunes of theoretical curiosity. A heterodox group, whose purported "schools" of thought, practice, method, and doctrines were transmitted in an oral tradition later interpreted and put to use in the interests of the writings of their doxographers, the primarily Miletian thinkers launched a tradition of unbridled theoretical inquiry regarding the phenomena of the natural world. They also provide a reconsideration of the very nature of the beginnings of knowledge that echo in strands of contemporary thought.

In antiquity's cradle of speculative inquiry into all things, the Presocratics were constituted by small communities and individuals who Ricoeur (2007) calls the "men of infinite tasks" through whom a "leap was made from the will-to-live to astonishment, from opinion to science" (p. 153). Rigorous disciplining and initiation were frequently pedagogic prerequisites. In the Academy and the Lyceum later it was a quest for and recognition of the intelligibility of another world beyond what appears in the bodies of living things and the ecology of situated minds. It was nostalgia for eternal forms, a "remembrance" of ideal, non-sensible and pure forms. Invisible, and non-tactile, it was also by Plato's time a mathematically perfect harmony that no one could hear, except perhaps the Pythagoreans idly playing their cythera.

The quality and limits of what is humanly knowable was a key point of contestation for the Presocatics. It remained a question among diverse doctrinal communities following the precepts of Xenophanes, Hecateus, Heraclitus, Parmenides, Thales, and Alcmaeon. At root, both an epistemological and ontological challenge was staked out from the Homeric age into the classical Hellenic Hera. It situates one of the most constant controversies in the practice of philosophy and science. It is at once a thinking of finitude and the nature of change in a quest for mastery of the universal and grounding principles of things. The Presocratics, to a considerable extent, appear to have managed to simultaneously cohabit and resist the dominant Olympian, Orphic, and Dionysian cults and myths. As Vernant (1982) explains, "sophia" and philosophia, in order to function as semiautonomous practices of truth, needed to substitute their own, newer ordeals for entrenched Orphic mystery cultures (p. 59). The mythoreligious and mathematical nature of the Pythagorean tradition, as well as the trains of thought propelled by the Miletians, including Thales, Heraclitus, Parmenides, Anaximander, and Anaxogoras, required providing a path of *askesis*, in ritual initiation practices. This disciplining of disciples preserved and appropriated ancient divinatory practices and spiritual exercises aimed at maintaining a specific structure to the economy of attention that Hadot finds functioning in the community surrounding Plotinus centuries later. These disciplining practices gave coherence, focus or *skopos*, to their distinctive pedagogical forays into theoretical curiosity.

The Presocratics also brought new methods of discussion and argumentation. Most germane for the discussions in following chapters, they anticipate a materialist investigative, if not particularly experimental, method of inquiry. In the case of the Pythagoreans, this included the novel innovation of mental heuristics and models of universal proportion, analogy, and ratio through the study and contemplation of forms provided by geometric and mathematical logics. For the ancient Greek world, the speculative and the practical, intellectual and manual labor, are set into a polarity that does not close off their synthesis and unity in difference. Theory as *episteme* is connected with its worldly counterpart practices. These practices are usually referred to as *techne*, *metis*, and *poiesis*—techne referring to productive, crafting knowledge; metis to the "cunning intelligence" necessary for the practical skills of hunting, fishing, or warfare; and poiesis, to the making new of a repetition in any art, whether rhetoric, theater, or art. These practices shared, appropriated, and supplemented traditional models and modes of thought, belief, and pedagogic transmission. The learning provided by the natural philosophers often included rejection of the lures of indolence and sensuality within an ideal of arduous effort or *ponos*, for the pursuit of the utility of knowledge in the interests of working toward a consistently ethical life.

In Anaximander we find a vivid personification of theoretical curiosity in the ancient world. His protean inquiries set into motion trains of thought still running and finding conjunction in contemporary research and thinking. For Kahn (1994) he "emerges more and more clearly as the central figure in 6th century thought. It is, in all probability, his work which laid down the lines along which ancient science was to develop and his mind which gave the Greek philosophy of nature its characteristic stamp." He created the literary genre of a usually abstractly argued inquiry into nature itself, the *peri physios historia*, and was the first to utilize prose in his works. His literary influence shaped the thinking and writing of those directly influenced by his inquiries into nature, including Plato's Timaeus, Aristotle's physical treatises, and Lucretius's *De Rerum Natura*. Kahn claims that Anaximander's commitment to the genre, to the act of writing his inquiries into nature "brought the history of Greek philosophy into existence and represents nothing less than the advent, in the West at any rate, of a rational outlook on the natural world" (p. 6).

Anaximander enters the dispute about the primary elements of nature. He proposed no primary element. No single substance was to be considered *arche*, the ground of all the things, originating the combinations that formed the material universe. In opposition to those who advocated for a primary constituent element (Thales's advocating that "all is water," Anaxemes's advancing air, Heraclitus's fire), Anaximander proposed that there was nothing that formed the ground with respect to the *apeiron,* the boundless or unlimited. Instead, the preexistent *apeiron* itself was arche for all the rest, encompassing and governing all the other processes and combinatory physical properties of natural things. Theoretical speculation regarding the elements is the first example of how inquiry, even those concerned with the first principles of the natural world, refracts and is already enmeshed in the social controversies and concerns of its time. Vernant reads Anaximander's principle within the political plurality intended for the Greek polis in that his premise of an infinite and immortal generative power required the reciprocity of all elements and processes. Vernant (1982) claims that he "delineated a new image of the world so rigorously that it formed a sort of common ground for the pre-Socratic philosophers as a whole" (p. 124). This is evident in medical thought as well, as in Alcmeon, for instance, who regarded health as a balance of powers.

A tension between the nature of the senses and universal principles is also evident in the movement toward abstraction in the Presocratic cultural context. In relation to Parmenides's maxim that "all things are one," Richard Seaford (2004) suggests there is strong cultural and material sources for this purely speculative, metaphysical abstraction in the increasing circulation of money in the form of minted coinage. This amounts to making the claim that this ancient transition to minted metal, as universal exchange, paralleled a secularization of a pre-monetary culture of reciprocity, ritual, and kinship. In Vernant's view, it manifests a secularization of the logos. The pure abstraction of an unchanging and timeless One acted as a projection of the substance of exchange value, of a commodification of the prior culture of gift exchange. For Seaford, Parmenides's One is not a "sudden break" with a gift and sacrificial, ritual society, but the culmination of a more gradual reorientation toward the abstract already present in Ionian cosmology. This was paralleled and to some degree augmented by the introduction of the minting of electrum and copper and its rapid circulation across and between regions. It became the measure for all things conceptual, allowing for an abstraction from materiality into that of the formal, ideal, and non-sensuous.

This historical analysis helps situate the speculative inquiry of Anaximander, whose "boundless" *apeiron* may have had a basis and analogue in the material cultural milieu of his time. It also indicates how a cosmic and Olympian mythopoetic culture of sacrifice, with their timeless, struggling gods, turned toward an impersonal principle of generation and measure for all things material and immaterial. The logos, the Parmenidean One, and Anaximander's "boundless" and "unlimited" emerge then as homologues of the equivalence and exchange relations set into social relations with the prevalence of coined, minted money. Aristotle inveighed against the unlimited accumulation of money as personified in the myth of Midas. Seaford (2009) insightfully notes that he cited this story to illustrate the arbitrariness of monetary signs as they function autonomously, unmoored from the material realities of everyday life. Seaford observes that the "Greeks of the classical period were anxious about the potentially unlimited scope and power of money, and this anxiety contributed to their explicit privileging of limit over the unlimited, especially but not only in metaphysics and ethics." He cites Plato's *Philebus* as one instance in which the unlimited is considered to be in need of control. Aristotle's *Nicomachean Ethics* also comments that the "bad is of the unlimited, as the Pythagoreans surmised, and good is of the limited" in what he claims is a "culture of the limit" (pp. 14–15). He also stresses a point pertinent to my central concerns regarding the shifting valences of theoretical curiosity. An unbridled speculative theoretical attitude existed, and as I argue in succeeding chapters, continues to live in degrees of opposition and tension with forms of thought that are regarded within their cultural frameworks as practical and utilitarian.

Distinctions between divine and human knowledge were the turning point on which much the Presocratic inquiries were either begun or kept in check. For Xenophanes the key contrast was between what could or could not be reliably known. Appearance only is manifest to humanity, as semblance, *dokos*, with no person capable of knowing *to saphes*, what is clear and indisputable truth. For Homer, the distinction had pivoted on exact knowledge of the eyewitness and that of hearsay, of deceptive, changeable, human knowledge: "truly the gods have not revealed to mortals all things from the beginning, by long seeking do men discover what is better" (Vernant, 1982, p. 127). For the Presocratics, what emerges is the frequent claim that through their own strivings for knowledge humans can acquire knowledge. This tenet, enabling the adventures and fortunes of theoretical curiosity, stands in stark opposition to an earlier period, as represented by Hesiod with an irreparable rupture between divine and human knowledge.

Only beginning with Xenophanes apparently is it human's initiative that knowledge of both the perceptual and supersensible become possible. It is through the intentional and deliberate exercise of human will that, for Xenophanes, the acquisition of knowledge develops. His precepts appear as a sign of a gradual and increasing secularization of language, poetry, and speech. Detienne (1999) identifies this secularization in the use of speech acts that come to replace "efficacious speech" forms, including the mantic invocation of the Homeric era of the muses and mystery cults (p. 17). It is a critical shift toward speech and rational discourse that would take form in the disputations, intellectual inquiry, and political development in the agora of the city. Logos, speech, and language would become more autonomous, and, rather than conveyors of timeless truth, *aletheia*, would move to the center of civic life.

Xenophanes claimed that our knowledge will always remain incomplete and obscure. While wisdom remains always the ultimate goal of human aspiration, our ability to know the ways of the world are only developed by constant search. He asserts the right to rational inquiry through observation, debate, and a discourse that, perhaps for the first time in the classical context, bridges the gulf between the gods and humanity. Hecateus, following Xenophanes, affirms more substantively the independence of human inquiry. Setting out to discover what is true, he claimed that knowledge consists of the data gained from sustained inquiry. Unlike Odysseus, whose search and journey are forced upon him unwillingly, Hecateus encourages well-organized and systematic planned journeys for the purpose of gathering observations of the natural world and the customs and deeds of the peoples of the world, including the very detailed and perhaps first attempt to make a representational map of the world, as attributed to Anaximander. Here Herodotus as the generator of *historie*—coherent accounts or inquiries, documented and observed cultures, customs, politics, war, and the wondrously strange—appears exemplary. Taking note that Solon, who, at his leisure was the first to travel the world for the sake of theory, may have helped legitimate his own lifelong collection of travel and commentary. For Hecateus, experience is the one and only basis for true knowledge, distinguishing what he himself sees, what he is informed of through eyewitnesses, and from rumor. Humanity becomes independent by setting out to discover what is true.

Throughout antiquity the historical place and purpose of *theoria* witnessed shifting semantic play and ethical value. It is pertinent to my purpose here in documenting some of its appropriations in distinct epistemic contexts. By the time of Xenophanes and Hecateus the meaning of theoretical projects returns and retains the value attached to the visual. Forms of perception once again are the pivot for arguments and anxiety about the relation between the sensible and the intelligible. Herodutus's actual and fabulous travelling is perhaps the first document that presents the presence of a "spectator" persona, a distanced perceiver, one who apprehends the myriad of the worldly objects of the senses.

Experimentation was attempted and considered of value, even within the limits of the instruments available to the ancient "doctors." Empirical practice was neither foreign nor entirely uncommon from the time of Anaximander up to Aristotle's formalization of natural scientific inquiry. The issue of empirical and experimental approaches is often raised about a classical world often regarded as prescientific in its methods of inquiry. The speculative interest of many of the experiments documented are the collections of evidence in support of a researcher's own theories. Hadot (2006) calls this a process of "saving the phenomena," or what, in juridical terms, would mean a leading of the witness (p. 163). The "natural doctors," concerned as they were with first principles, saw little need, and had few instruments by which to ascertain the regularity and patternings of phenomena. They sought corroborating evidence for their usually epistemic claims, rather than heuristic models for the continuity of what we now identify as research programs. A strong empiricist strain clearly existed throughout Greek antiquity. Though induction was deemed inferior to their rationalist, epistemic, and speculative interlocutors, empiricist practitioners, like that of the Miletians, skillfully studied the movements of the stars, winds, and natural phenomena within the parameters of available observational techniques. Brunscwicq (2000) details a broad range of empirical practices, including the principle of *autopsia*, or direct and personal vision, clinical and case studies (as we find in Alcmaeon and the Hippocratics), the general *historia*, recollections and testimonies of past and present events, as in Herodotus, and *metabasis*, or inferencing and transfer of accepted theses to additional and plausible knowledge of the yet to be disclosed (p. 28). More experimentation was apparently possible in their wide-ranging forays into physics and biology. There were skilled observers dating back to Thales's astronomic predictions, Anaximander's maps, as well as attention to detailed analysis by Democritus that Aristotle praised and emulated.

Criticism of the experimental naivete of the ancient are misdirected, For G. E. R. Lloyd (1970), experimentation was not especially relevant to the kinds of work pursued in Greek antiquity. It is not a question of a lack of a prescientific precursor to the experimental methods of modernity. It is the degree to which experimentation was relevant to the foundational problems and questions they posed, and the extent to which these were amenable to empirical method. The first requirement for these researchers was a need to attain clarity of fundamental epistemological problems that concerned them collectively before initiating any sustained observation of empirical evidence. One of their primary concerns, deriving from Parmenides, remained centered on the question of *physis*, of phenomenal change in nature. For well over four centuries questioning the elemental substance, quality or substratum of matter, was vigorously pursued. This core question framed an abiding theoretical interest, setting into place diverse models both for the "natural" kinds of change apparent to the senses and explanatory frameworks for the genesis of the world.

Astrology, the paradigm of inquiry and meteorology, necessarily remained largely speculative without lenses or mirrors, as were the perennial debates in ancient physics about the "ultimate constituents of matter." From Anaximander on, familiar phenomena provided the basis for logical argumentation, often formally dialectic and through the use of analogies. As Lloyd illustrates, the controversies between the atomists and Aristotle could never be resolved by either observation or experiment but to assert "that they never experimented is patently untrue" (p. 140). Even as early as the fifth century, the Pythagoreans were making their esoteric discoveries in acoustics. For the Hippocratic writers there were studies of the freezing of water into the "bright, light sweet" that separates from the "muddy and heavy." It was not until late antiquity that more exacting and systematic inquiries were initiated. Galen's careful detection of aspects of the nervous system and Ptolemy's optical records exemplify a newly invigorated empirical experimental methodology. Concerted attempts were designed in order to discern causal relations of observable phenomena. Dialectical argument, syllogism, and analogies continued, however, as forms of immaterial mastery and truth.

In his formidably influential corpus of work, Aristotle embarked on a synthesis and reinterpretation of his predecessors. His reading of the Ionian "natural doctors," as well as of Parmenides and Heraclitus, would leave its impression, surviving through Latin translations from the Arabic in the middle ages, on the origins of modern scientific inquiry into the principles and processes of nature and life. His vast scope engaged its own questioning of the elemental principles and substance of things. The Presocratics, despite their quite distinct formulations, shared an abiding theoretical interest into the causes of things in both the natural and physical world. It is with Aristotle's refutations and formalizing of thought into "first things" that a complexly textured interest into the nature of life begins to cohere. In the systematic inquiries in the "natural books" he established the basis for a scrutiny of the principles governing what came to be known as natural philosophy. In formalizing a tripartite, hierarchical order for study, one that would serve as a precedent for the trivium and curricula of the early universities, he placed metaphysics or theology as primary. For Grant (2007) this is the domain of spiritual substances independent of matter, the unchanging and unmoving order of the world. The secondary domain was that of mathematics, in which one studied abstractions without bodies. The third is that of natural science or physics, the study of which entails analysis of all that undergoes change and that have an inherent dynamic power of motion. Grant tells us that the study of physics or natural philosophy for Aristotle was inclusive of "both animate and animate bodies and is applicable to the whole physical world, that is, to both the terrestial and celestial regions" (p. 39). Unlike Pythagoreans and Platonists he also importantly held matter to be indispensable, positing its function in nature as the material cause. Matter was to be regarded, though rather ambiguously, as the underlying and nameless substratum of all qualitative change.

Aristotle did not deny the value of sense perception. He understood the senses as the gateway to move from the particulars to the universal principles of the whole in any inquiry. In the *Posterior Analytics* he advances his inductive method, the *epagoge,* which required moving beyond the individual, qualitative, and quantitative aspects of a phenomenon. This method is one in which "one arrives at universals, from sense perceptions by means of induction. It is by means of universals, not direct perception, that we can generate demonstrations that produce scientific knowledge. As Aristotle explains, perception is of the individual or particular, and from particulars "it is impossible to perceive what is universal and holds in every case" (in Grant, 2007, p. 39). His discerning analytic, observational skills, which remained non-experimental, did include occasional interventions. He conducted some dissections, as in his observations of the embryological development of chicks of a hen. Grant says that Aristotle "obviously broke open eggs that had been laid at the same time and observed the status of the chick at different stages of its development" (p. 48).

Experiments were apparently considered to be of little or no value in relation to the meticulous observations that would come to comprise his inductive method. Grant views Aristotle as reluctant to intervene into natural processes in that he could find no benefit in interfering and interrupting the processes of *physis* that were so central to his life work. It is also possible that even he "was not ingenious enough to conjure up appropriate experiments that might shed light on the natural phenomena in which he was interested. It is not at all clear that Aristotle was reluctant to intervene in nature. A more likely conjecture might be that he rarely ever thought he had to, because he was convinced that he could derive solutions to most problems by contemplating the way things had to be by a priori and deductive means" (p. 5).

The question of the social and cultural relations that theoretical curiosity participates in will arise in the context of each of the following chapters. In particular, I will be interested in a phenomenologically accented concern for the place and appearances of physis and techne in modern technoculture. This is the question of the role of theory within an era that witnesses the dominance of technical forces and forms of reproduction. The pedagogic and ethical quality of the relations between practitioners and their material culture will be especially germane in my concern in Chapter 6 for new relations, or a "writing" of novel forms of life are now made actual by contemporary research under the auspices of the genetic and genomic technosciences.

The conditions of research are also central to this project. According to Diogenes "Lives of Philosophers," Heraclitus, Parmenides, Anaximander, and Empedocles were born to wealth and property. Being raised in artisanal or craft families seems to have been the exception. Even Thales, the prototype of speculative inquiry, the unwordly philosopher, is in some accounts a quite cunning man of commerce; in his business acumen he appears to have held a corner on olive presses, and to have engineered diverting the Halys River for purposes of trade. The economic privilege of many of the original *physikoi* can be taken into account as one of the decisive factors for their engagement with the currents of thought in their time, as well as their position to influence and manage practical decisions of governmental, military, and engineering importance. For my purposes, it invites a questioning of the interests of theoretical inquiry. Speculative theory has, since the classical Hellenic age, often cloaked its own partisan commitments, including its class interests and implicit agendas, behind a veneer of "pure" and disinterested research. Lloyd (1970) remarks that Aristotle's life and corpus evinced an "incredible complacency concerning the material conditions of his day" (p. 132). In the context of antiquity, raising the issue of privilege can also demystify their own debates regarding the core question of whether knowledge and inquiry can or must be useful. Lloyd reads the primary thinkers of this time "sufficiently free from immediate financial worries to be able to spend a deal of time and energy on a wide range of relatively unproductive, even counter-productive activities" (p. 126).

Institutional support or indifference figures significantly in the various crosscultural fortunes of theoretical curiosity. Lloyd and Sivin (2002) document some of the component alliances in classical Greek and Roman contexts and compare them with the quality of the structures of support, their "cultural manifold," in ancient Chinese culture in which [4]"issues of state control, or its subversion, could be at stake" (p. 13) in the maintenance of schools of thought and their individual practitioners. They cite as an example that a casting of the horoscope of the Holy Roman Emperor was deemed a form of high treason. In Greece "no early Greek historian had any prospect of employment," requiring considerable skills in self-promotion in the essentially competitive and individualist knowledge work of rhetorician, orators, lecturers, who like, Herodutus, made a living reading their work in the public squares of Athens (Lloyd, 2002, p. 17). Both the Academy and Lyceum, begun in the fourth century BCE, with no instruments or mechanisms to maintain, were self-supporting, the former excelling in mathematics and its pedagogy, and the Lyceum in biology. Teachers found their livelihood and collaboration in research. With the opening of the museum in Alexandria in 280 BCE, research comes to depend on structures of patronage.

The analysis of differing institutional contexts of research and inquiry raises the specific question: in what manner do institutional frameworks contribute to the "factors that stimulated or inhibited systematic inquiry" (Lloyd, 2002, p. 126). Examining individual proponents, Lloyd observes the contingent "fortunes of transmission," of the likes of Heraclitus, or the third century BCE Chinese thinker Gongsun Long who gained notoriety for his paradoxical argument concerned with proving that a white horse is not a horse. Outside of those who gained recognition through their teachings and the peripatetic work required of them, unknown numbers of researchers and their inquiries were consigned to obscurity. In Greek culture, in contrast to China, institutional support such as that provided during the reign of the Ptolemies was the exception to the rule of individuals competing for resources and recognition. The Ptolemies founded institutes of research and documentation, including the museum and library at Alexandria. They also actively sponsored research outside of the library and museum by providing opportunities for anatomical research on human subjects. For the most part, however, "there came to be salaried positions for a very limited number of those who taught Platonic, and other, philosophies at Athens (that is, basically the heads of various schools)" (Lloyd, 2002, p. 129).

Theoretical curiosity is frequently conflated with a more mundane disposition to distraction and novelty, to what some twentieth-century experimental psychologists (as discussed in Chapter 3) classified as a "perceptual" or "dispersive" curiosity. In the late classical and medieval era it was scrutinized as a spontaneous unrest, as dissatisfaction, a being driven about, a *polypragmonein*, busyness, and meddling in other's affairs.[5] An epistemic shift of importance occurred with the transition into the classical, Hellenic era. It was manifested in a general cultural "downgrading of perception." It appears to be at work in a cultural milieu in which the transition and relations between theory and practice remained in flux until the time of Plato's Academy (Lloyd, 2000). It will be with Aristotle's reinterpretations that a consolidation of the divergent trends of the Presocratics coalesces into what will be a canon for generations. This ambivalent defamation of the senses and sense knowledge will remain a prevailing cultural disposition throughout much of the ancient and early modern periods. It will sanction the legitimacy of a purely intellectual knowledge, shaping the permissible forms, styles, and contents of inquiry. Speculative theoretical curiosity ascends to a status that will structure research into Western modernity. Curiosity is derided, associated with the narrative historical accounts like that of Herodotus. It is relegated to the inferior realm of appearance and ephemera, of processes of becoming, change and motion. Only a speculative gaze inward and "upward" is prized and granted value. A renewed interest and mediation between the intelligible and the sensible is retrieved under the influence only later with Neoplatonist practices, which came to the fore in the early modern period.

A dilution of wonder, of the originating wonder, *thauma*, the spark to curiosity that Aristotle remarks upon, had already begun in his own time. The eclipse of sensory knowledge was perhaps partially an effect of his own vast and misappropriated historical influence on the processes, methods, and limits of inquiry. This pervasive "degrading" of sense knowledge, as in the loss of a phosphorescent awe at the heavens by early astronomers like Anaximander or Eudoxus and Aristarchus, is an ambiguous, contradictory process. It has taken place in multiple "Western" contexts, varying in the extent to which empirical and observational research could be meaningfully conducted. In the historical context of inquiries in the Presocratic and the later, Athenian "schools," with the important exceptions of the Epicureans and the atomists perhaps, it performs a structuring and disciplining of the senses. It marked a marshalling of a logic and discourse in resistance to the evidence of the sensible, a defeating of perception's lures and chimerical inconstancy. The principles of the world which might at most be abstracted primarily from vision, but also of touch, as for Democritean materialists, were regarded as autonomous, eternal, and ideal forms of intelligibility.

It is important for this project to stress that the senses came to be considered during the Hellenic age as inferior to an intellectual grasp of the nature of things. This sundering of the knowledge contributed in large measure to the classical legacy as one resisting human sensuous embodiment. For my consideration of the phenomenal and experiential pivot and ground for all intellectual discovery and theoretical practice, its significance cannot be underestimated. The world of appearances was phantasmal, conditioned imitative images of what was beyond sensory and perceptual recognition. Things, animals, plants, birds, and civilizations could be regarded and discarded as ephemeral, insubstantial aspects of the phantasmorgic realm of continual change and becoming.

It is this purely speculative, disinterested curiosity that I oppose in advocating for critical and materialist modes of inquiry. I consider the Pythagorean and Platonic models identified here as theoreticist. They are beacons for an abstracting rationality. It was the theological and cosmic contemplative state that Presocratics like Xenophanes attempted to reject in favor of what could be known and fashioned by finite mortals. It is also a basis for the Neoplatonist influences in melding of sources that augured the eventual emergence of the Christian faith. I wish to make this crucial distinction between an engaged, critical theoretical practice and that often connoted by an immaterial speculating theoreticism in the following chapters. The contrast between speculative theory making, for an ahistorical "spirit of system" for its own sake with the potentiality at work in a robustly theoretical curiosity is the challenge of the passion of the mind and body. This informs my attempt to glean some of the significant points of departure, persona, and events that have called for sustained inquiry.

Almost unanimously among the Eleatics, including Parmenides, Zeno, and Melissus, it is the evidence of the senses, especially the powers of sight, that become the object of uncertainty and reflection. For thinkers like Empedocles and Democritus the senses yield only a bastard knowledge of shifting and fleeting phenomena. This premise led to advocacy of a moral marshalling of attention and the passions. From Cicero to St. Augustine curiosity becomes a sign of a human existence losing its essential centering, one that seeks to be satisfied from outside, by external things. Human existence has become blindsided. It is associated with those who have forgotten their origins and relation to what is spiritual, having always already fallen into what amounts to a moral and ontological dispersion. The issue of the gaze and attention returns in the recurrent condemnation of practices causing this fallen condition, frequently considered in antiquity to be brought about by seeing the many instead of the one.

Xenophanes and Hecateus here stand in contrast to the august legacy of Heraclitus for whom the philosophic individual, though he must make "inquiry into many things," prioritizes speculation and a concern for objects which cannot be perceived. His gnomic fragments attest to a search for visible signs that would work as a means for the attainment to the realm of the invisible through which the wise catch glimpses of the secrets of life. What remains invisible, accessible solely through a purely theoretical pursuit of the unifying principles, is truth. The invisible nature of things, including physis, exists outside the experience of humanity, partaking of a thoroughly divine knowledge and power.

The question of visibility remains a key turning point in the fraught debates between the schools of thought in the Presocratic and classical world. Alcmaeon, a disciple of Pythagoras, taught that nothing more than conjecture is possible for a humanity that was limited by the range of perception. Maintaining the distinction between divine and human knowledge and its analogous relation of invisibility and visibility, Alcmaeon, who is one of the first to apply natural philosophy to the human body as the basis for medical practice, sustained the influential notion dating from Homer and Xenophanes that to know can only mean what sight provides. The visual is here set in counterpoint to much of both the classical and modernist values; it serves as our primary access to knowledge of the world, of our bodies and phenomena.

It is impossible for humans to know what is distinct and clear, *to saphes*, the underlying and immutable truths, the logos, and laws of the universe. What is distinctive about Alcmaeon's teachings is their overriding attention to the non-evident, or not yet evident. Humans, in their striving to approach the invisible through observation and conjecture, are able to make reasonable conclusions from clear indications of recurrent and patterned experiential phenomena. He distinguished three kinds of knowledge: first, those of beasts, ruled only by sense impressions, in their grasping of appearance; second, divine knowledge of the invisible; and third, singularly human knowledge capable of combining sense perceptions and making informed predictions and prognoses about the non-evident and invisible. Human knowledge is distinctive in its utilization of sense impressions through a working through and synthesis of the given. The function of the brain is to transmit sense perceptions, which in turn allow for recollection and assumption. Only after finding degrees of certainty in these can contingent claims and tenets of knowledge be established. His precepts praise the value of working through sense experience in approaching the invisible, not by intuition, as in Heraclitus, but through an orderly, rigorous method of analysis. His criteria were physiological and psychological, expressing an empiricist approach that would carry on through the work of Empedocles and Hippocratus.

With Socrates, it is the concern for human affairs, not the *physis* or cosmos of the natural doctors, which is paramount. His method expresses a secularization of the logos. In the *Apology* his knowledge is only human—those of others who speculate on problems of natural philosophy were attempting possession of a superhuman knowledge. He rejects the myths and fabulations of the Homeric poets, seekers of divine or natural knowledge, vainly attempting to attain to the truths through the channels of human understanding. Socrates, as personified by Plato, who attended his trial, appears as a decisive turning point from earlier investigations and practices. It is a break that Nietzsche (2001) will mark as a passage between Pre-Platonic, not Presocratic, systems of inquiry. While primary, motionless, and eternal forms remain central to the Platonic structure of knowledge, transcending the sensory world of images and simulacra, an erotics also motivates a disciplined pursuit of virtue. In the *Symposium*, the sibylline Diotima, as recounted by Socrates at a hearty bout of drinking, teaches him how love mediates between the terrestrial and the divine. Acting as "messengers" between the ephemeral and the permanent, the transitory and the universal, erotic energies fill the seeker, in becoming a lover of the universal nature and truth realized through a "birth in beauty" (Plato, 1997a, p. 493 [206e]).

Socrates's famous method, his midwifery of thought (ironically, perhaps, referencing his own mother, Phaenarete's humble status and occupation), was one requiring a perpetual process that required to "live the life of a philosopher, to examine myself and others" (Plato, 1997c, p. 27 [29a]). The contingency of truth claims would always be subject to the rigors of a questioning, seeking "universal definitions" that challenged the foundations of all existing presumptions to secure knowledge. He was a paragon (or nemesis) of engagement with the everyday affairs of the life of the city, which ultimately brought him to trial in charges of corrupting the young, propagating his practice in what was taken up as fundamental challenge to doxa and the gods of the city. Socrates reflected the reality of a life, as Gregory Vlastos (1991) points out, largely spent as an "habitue of the marketplace, missionary to the unwashed" (p. 251). He remained to the end, on his own terms, loyal and devoted to Athens and the furtherance of the good life it made possible.

The place of wisdom, *sophia*, becomes foremost critically attentive to the problems of the human world, the polis and the politics of a contingent historical place. *Sophia* and doxa, common belief, are forms of human knowledge that alter in relation to the changing dynamics of a particular era. Sophia redirects attention away from the natural world, the universe of *physis* dominating much of the speculative, protomaterialist inquiries of the Milietians and other earlier investigators and "schools" toward the pressing and enduring concerns of human existence. Inquiry centers on the the elements of social life, of how it is made up by the forces that could divide it against itself, and the means by which they might be harmonized so that their conflict might give birth "to the human order of the city" (Vernant, 1982, p. 40).

In the emergence of the Academies during the Hellenistic era, the cognitive drive asserts itself. The truthseekers or *mathematokoi* resist the knowledge of the senses and of "hearsay." They aspired to transcend the limits of perceptual appearance, simulacra, and change. There emerges a collective drive toward a mathematization and systemic ordering and limiting of knowledge and its internal relations. The seekers of truth sought abstract, dematerialized principles. A pedagogy of "pure" theoretical curiosity laid claim to universals outside of time or place. These principles were the generators of the principles by which civil life and politics would be governed. This intellectual pursuit, which would reemerge in the early Renaissance among Neoplatonists, coexisted paradoxically with a general cultural appreciation and dependence on the arts of *techne* and *metis*, the practical knowledges of commerce, trade, navigation, and artisanal craft. An intensifying rigor of a mathematical and geometric model of knowledge pursued "pure thought," virtue and pure theory. Wilbur Knorr (2000) makes the provocative claim that the "transition from praxis to *theoria*—from mathematics as techne (techniques and methods and designs within practical, material activities) to mathematics as episteme and gnosis (a form of pure knowledge)—occurred only once in human history, namely among the classical Greeks" (p. 235). Though it may have occurred at other times and places, a significant change in the place and purpose of theory certainly emerged during this time. It has created a legacy that continues to influence contemporary thought and research practices.

By the time of Plato's *Republic*, this purported epistemic shift was markedly present. Its paradigm was an abstract, deductive logic that ideally shaped the structure of dialectic and rhetoric alike. It was premised on the demand and prestige for demonstration, as manifest in the developments of geometry from the Pythagoreans to Euclid, one that is increasingly, if not entirely, divorced from the techne of seafaring, construction, and practical measurements. The earliest geometry, including the practical applications of Anaximander, may have begun among Egyptians surveyors in attempting to reestablish control over land in response to the regular flooding of the Nile. Thales himself, the object of ridicule in Aristotle's account, acted in the role of what we might now call a military attache, influencing the outcome of a battle with the Persians, by his accurate prediction of a solar eclipse. A concern for numerical precision may have developed in navigational and trade relations begun among the Phoenicians who also provided the impetus for the dissemination of what crystallized into the Greek alphabet. The Neoplatonist Proclus justified this paradigmatic change, asserting that "it is not at all surprising that the discovery both of this geometry and the other sciences (*epistemai*) commenced from need, since everything in genesis proceeds by moving from the incomplete to the complete (*teleion*). It is thus reasonable that the transformation would occur from perception (*aesthesis*) to reckoning (*logismos)* and from this to mind (*noos*)" (Proclus in Knorr, 2000, p. 236). In this perspective, mathematics in general, and the "project of geometry is not the manipulation of figures in physical constructions but the understanding of their properties in pure thought" (Knorr, 2000, p. 235). The shift to a more emphatic stress on the powers of deductive reasoning may have originated among the Pythagoreans and their premise, as recorded and interpreted in Aristotle's *Metaphysics,* that all things are numbers. Geometry was paradigmatic in that

> they thought that the principles of mathematical entities were the principles of all entities. And since of mathematical entities numbers are by nature primary, among these they seemed to observe many similarities with entities and things coming into being, rather than in fire and earth and water. . .Since then all other things seemed to be assimilable to numbers in their nature, and the numbers were primary of the whole of nature, they assumed that the elements of the numbers were the elements of things as a whole, and they thought that the whole of heaven was a harmony and a number (986a, p. 20).

Plato, who according to Proclus, was "everywhere to awaken wonder about this subject of mathematics among students of philosophy," was a contemporary of the geometers Archytas, Theaetetus, and Leodamas (Knorr, p. 247). A foundational shift, this speculative, deductive *theoria* and intellectual curiosity was severed from the realm of the senses and embodied experience. It defined the epistemological pursuits of succeeding generations. Centuries later, one with no less stature than Galileo would proclaim that the "Book of Nature is written in geometrical characters" (Sohn-Rethel, 1977, p. 126). The abstract and the ideal were the vineyards and gardens of the Academies, their *scientia* and *matheme*, figures of the symmetry of the absolute. Plato sits in counsel with later-day Pythagoreans, who listen to silent harmonies to tune their spirit harps to a universe of number and proportion. Theaetetus, also a geometer, sits with Socrates, considering timeless forms and the modes of knowing immaterial things.

The ambiguity and contradictions of theoretical curiosity's place arise from its adventures in cultural logics. It has always been imbued with the abiding tensions and contradictions in the organization of daily life. It is inseparable from the textures of theory and practice, intellectual and manual labor which continuously compose it. The centrality of mathematical and deductive reasoning to theoretical curiosity deserves emphasis. It has served as an often-unremarked motivating spur to the will to power in successive historical regimes of truth. In classical, medieval, modernist, and contemporary epistemic structures, divergent systems of mathematical rationality have aided in the generation of paradigmatic and institutional logics of inquiry. This is evident in the mathematization and mechanization of nature that were later primary to the new empirical sciences. These models mobilized the unprecedented technical forces invigorating nascent capitalism's productivity in its "primitive" accumulation of natural and human resources. The speculative and "pure" theoretical curiosity of mathematics, natural science, and physics are the fulcrum for a particular ordering of knowledge and its means of production. They have naturalized their dematerialized ways of knowing. A disembodied speculative spirit has granting an ahistorical ground to explore and exploit a totality subject to unlimited research

I find a contemporary reprise of the premises Richard Seaford (2009) makes about the emerging primacy of money as a homogenizing and abstracting force in classical thought and social relations. Ancient Greek culture was transformed by its monetization, and modernity's global capitalism exponentially displaces cultural practices as an unchecked force for unlimited accumulation and the calculability of all things. Theirs may have been predominantly a "culture of the limit," but ours is what Seaford regards as one emphatically opposed to closure. This "hostility to closure" or limit "is evident in various spheres: economic, metaphysical, conceptual, narrative, and others" (p. 15). Anticipating my profiles in succeeding chapters, this orientation has become entrenched as hostility to any ethical and democratic closure to speculative theoretical discourse, including academic capitalism. The mathematization of social relations structures distinct historical hierarchies within and between aspects of manual and intellectual labor, techne, and praxis. It manifests what Sohn-Rethel, regarding the rise of modern science, as discussed in the next chapter, documents as a relentless evisceration of the material, physical aspects of work by the actual producers of wealth and knowledge. It is the signature of the abstracting rationality of commodity exchange relations inherent to the functioning of capitalism.

The Fall of Theory

Curiosity emerges with a dramatically different valence in late antiquity. It is charged as a vice by the early church fathers. In his treatise *de Curiositae*, Plutarch pronounces its vociferous denunciation. He identifies inquiry with gossip and an inherent wickedness of interest in others. As one of his rhetorical, moral expositions, this text charges the curious with a pernicious vice. He admonishes those who would gossip and take interest in the affairs of other men to "block up the windows and side-doors of your curiosity that open on your neighbor's property and open up others leading to your own" (Plutarch, 1939, p. 475). This straying of the mind and attention is a

> malady, a desire to learn the troubles of others, a disease which is thought to be free from neither envy nor from malice. It creeps in, searching out with slanderous intent, drunken revels and all-night festivals" so the mind of the busybody is at the same time in mansions of the rich, in hovels of the poor, in royal courts, and in bridal chambers of the newlywed. He searches out everybody's business, that of strangers and that of rulers, nor is this search of his without danger; so those who search out the vices of those more powerful than themselves destroy themselves before they acquire their knowledge (p. 483).

Plutarch is excoriating the lives of the *polypragmonein*, the busybodies and meddlers. It is clearly not the realm of inquiry into the nature of things, of *physis* or the universal laws and patterns of the world. He declaims

> what escape is there, then, from this vice? By a process of shifting and diverting our inquisitiveness, as has been said, and if possible, by turning the soul to better and more pleasant subjects. Direct your curiosity to heavenly things and things on earth, in the air, in the sea. Are you by nature fond of small or of great spectacles? If of great ones, apply your curiosity to the sun: where does it set and whence does it rise? Inquire into the changes in the moon, as you would into those of a human being: what becomes of all the light she has spent and from what source did she regain it? Nature is not vexed with those who find them (secrets) out. She is also like some plants "always blooming and green and rejoicing in the display of their wealth" while others like a human spendthrift, squander all at once their abundance" but perhaps you will have no curiosity about these subjects since there is nothing evil in them. Yet if your zest for meddling must by all means be forever feeding and dwelling on depraved things, like a maggot on dead matter, let us escort it to history and supply it with an unstinted abundance of evils. For there you will find "the deaths of men, the shufflings off of life," seductions of women, assaults of slaves, slanders of friends, compounding of poisons, envies, jealousies, shipwrecks of households, overthrow of empires. Glut and enjoy yourself (p. 485).

The cure for the life of the meddler is a concerted return of interest toward physis, as can be known in observing the heavens, earth, air, and sea. He declares that nature has nothing against this "cognitive drive" or appetite itself. Despite his eloquent denunciations, his interest, continued in what we read in early Patristic treatises and doctrines, is in the directing of attention toward appropriate objects of study.

Shorn of its association with everyday distractions in general, and their ocular, speculative fascinations in particular, curiosity carries a peculiarly ambiguous cargo into late antiquity. It is a signifier that refers at the same time to Plutarch's *polypragmeinen*, a meddling, prurient and morbid interest in the affairs and fortunes of others, and to the most sublime, abstract aspects of human intellectual and theoretical pursuit. As a term also ambiguously referencing the labor of the intellect, the creative detection of pattern and designs, it bears the imprint of the powers and faculties of the mind and the radical imagination. This ambiguous status is apparent in Plutarch's contemporary Seneca, whose *Natural Questions* was to have an immense influence on ideas of progress and discovery for the experimental attitude of such forerunners of science including Roger Bacon. Seneca advocated a passionate interest into the secrets of nature, claiming both that truths are owned by no one and "there will come a time when our ignorance of such obvious facts will amaze posterity" (Hadot, 2006, p. 172). He advanced research as an active, rather than contemplative generational and cumulative process. Antiquity provided the open routes for new research to be conducted on the part of and for the whole of humanity. Seneca affirmed that "the world would be a puny thing indeed if it gave the whole of humanity nothing to search for" (p. 173).

As discussed in the previous section, in the Presocratic and classical contexts, the work of inquiry is often speculative and conjectural. There is, however, a span of interest and concern for what we now consider to emerge as the empirical, as in Herodotus' *historia*, in contrast to the generally non-empirical natural sciences, or the disciplined, often ritualized practices of the *mathematokoi.* The epistemological speculations and scholarship it generates identifies membership among an elect "spiritual animal kingdom" Hegel will later describe, whose pursuits will transport their privileged adepts outside of time.

Contemporary researchers might naturally consider their practices those of a younger generation, freed of the sediments and obscurity of "folk science." As Hadot suggests, perhaps the ancients were the young ones. In their open histories of the human and natural world, they sought the principles of nature, of *physis*, of the genetic change within its creatures, the substance of the universe or the plurality of worlds, and of excellence, *aperte*, as the contemplation and purely theoretical curiosity in relation to the order of all things. Who is to say which age is younger or older, and whose questions are paramount?

Attending Medieval Minds

Early church fathers admonished against the free reign of the powers of imagination, warding off what they considered the wanderings of the mind. Christian commentators adopted the precepts and controversies of the late classical world, particularly the perceived need for directing of attention. These prescribed practices drew attention toward spiritual dimensions through specific, intensive exercises and practice. For Iraneus and contemporary Christian critics of Gnosticism, there were "rational" self-restrictions on an inherently unruly and disruptive cognitive drive. Conceived in terms of a relinquishing on the part of a ceaseless human inquiry, as exemplified by the existing Gnostic schools, restraints on young adepts and scholars to resist it were premised on the close association between an unbridled curiosity and vanity. The cognitive appetite was regarded as a pretension to emulate, or even gain some form of access to divine majesty. A theological model set its seal over knowledge of the physical world. In Blumenberg's account, the "hypertrophy of theoretical accomplishment is a new historical experience" (Blumenberg, 1985, p. 293).

During the first century after Christ, Clement declared that "he who goes beyond the region of what lies open to man and seeks what is withdrawn from him(what has not been plainly said or placed before our eyes) is one who prefers seeking to finding" (p. 295). For Clement, like Plutarch, it was essential to avoid a meddling in the affairs of others. It was necessary to reorient interest and to discipline oneself with the various sciences as a preparatory exercise. Curiosity could only function in the service of the truth, wisdom, a "second duty" (p. 298).

With no pedigree or aristocratic forebears, curiosity is at once a prodigious and prodigal energy, an orphan of the mind in its mature pursuits of truth and power. Often vilified in the scholastic communities of the Middle Ages, St. Augustine and patristic doctrines praise the conservation engaged in studied attention, *studiositas,* as opposed to the centrifugal energy of *curiositas.* Those "daring to know" posed challenges to key values, questioning clerical authority. Marked as a seducer by the church fathers, and castigated as a "lust of the eyes" by Augustine, *curiositas* represented both the road and messenger that divert the pilgrim on their journey to the holy land. It became the force of an often-demonic power to distract and confound the genuine scholar and student of scripture.

Ambrosius (337–397 AD), the Bishop of Milan, rejected Cicero's admitting of astronomy and geometry to the catalog of things worth knowing. Excluding these two disciplines, the centrality of human salvation retreat kept its unique and absolute preeminence. This reflected a reordering of man's spheres of interest (including the later Scholastic curriculum), breaking the ancient ascription of divinity to the stars. Ambrosius, who was influenced by Cicero's moral precepts, persuaded Emperor Gratian to banish all forms of heresy and was a compelling influence on Augustine's conversion. Philo's *De Migratione Abrhami* worked as a predecessor to Augustine's condemnation of curiositas with Abraham's wandering tracing out an allegoric path from self-estrangement to self-realization. The region of Chaldea signified the land of curiosity (*periergia*) and, specifically in the form of astronomy, remained a strong source of authority for Neoplatonist doctrine. Curiosity was understood as the false independence of the sensory aspects of conscious life, those which evaded their subordination to the intellect, and thereby diverting humans from the paths of deliverance. Abraham's legendary sojourn in Haran represented for Philo a turning from astronomical curiosity to self-knowledge and the fulfillment of God's work. For Philo, self-knowledge is the antidote to *curiositas*. He admonishes the "Chaldeanizers" to cease desiring astronomical knowledge, exhorting them to find shelter in themselves. In Blumenberg's summary of the era, an ethos of searching is newly deflected, as "things no longer show themselves but are shown" (p. 286).

With Tertullian (160–230 AD), the cognitive appetite was fully transfigured into a constituent element in the catalog of vices. It emboldened a Latin patristic literature in its attacks on Gnosticism, as well as directing specific fire against the "pagan" author Apulieus's apparent celebration of the unbridled curiosity by Lucius, his hero. His curiosity is like that of animals, credulous and capable of any indecency, including magic. He symbolized those who desire to know everything. Tertullian held that Gnosticism was rooted in an obstinate insistence of continual questioning. He attacked Gnostic teachings for permitting a pluralism of opinion, opening up divergent interpretations of the injunction "seek and you shall find." For this church father, there must be a prohibition against worldly searching and questioning. In a significant way his pronouncements announced a moral and epistemological closure. It was a decisive shift responding to an institutional settling and territorialization of doctrinal authority and position. Tertullian promulgated the decisive maxim: *"nobis curiositate opus non est post Christum"* –"after Christ, we have no need of curiosity" (p. 301).

Mindful of the lure of images and the phantoms of the imagination on the lives of novitiates in the monastic and scholastic communities, Christian doctrine made use not only of the fall of man in Paradise, but also adapted the pagan Greek myth of Pandora. The conflict and distinction between seeing—Augustine's ocular lust—and the struggle for knowledge of the Word of God through intensive reading of the Gospels and patrician commentaries remained a central concern in disciplining these communities. Early medievalists drew a parallel between Eve and Pandora, the latter incarnating not only the temptations of the flesh in the form of the first woman and her descendants, but also as a figure for unrestrained and dangerous epistemological zeal. Akin to an Eve irresistibly drawn to the forbidden fruit, Pandora gazes into the vessel that was fashioned for her by vengeful Olympians, releasing into the world the seven deadly sins. Pandora and Eve's vice is neither that of looking or eating the object of desire, but what Laura Mulvey (1996), in a contemporary cultural context, calls an all-consuming "epistemophilia."

A minor and formidable combatant, curiosity is arrayed in the battles of the passions. In medieval iconography, textual and architectural alike, it is depicted in animal forms. It can be found inscribed in iconographic tradition dating to at least the fifth century *Psychomachia* of Prudentius. The *Psychomachia,* or "soul struggle," depicts the classical virtues and vices engaged in combat. As a rendering of classical and Neoplatonist values and tropes, it set into dramatic, metrical relief the contest of the passions (Katznellenbogen, 1939).[6] For Bernard of Clairvaux, a passionate iconoclast and the nemesis of the all too inquisitive and independent Peter Abelard, curiosity is caricatured as a frog-eyed, big-eared monk. Christian Zacher (1976) suggests that the depiction of curiosity as an ape on the north facade portal of Chartres Cathedral indicates a continued fascination with an animal that "signified sensual pleasure and was itself an object of idle pleasure" (p. 166). In all its images, *curiositas* persists within the perennial or eternal struggle of the passions. As Hans Blumenberg claims, the desire for knowledge, *cupiditas scientiae,* was identified with the temptations spurring the first humans in paradise to "seek equality with God" (p. 326).

For church fathers like John Cassian (360–435 AD) one of the most powerful threats to disciplined study and the practice of monastic meditation was posed by pagan literature. The task of regimented *studiositas* was at once a learned, intentional forgetfulness of the baneful influence of non-canonical texts and a concentration on the purpose, *skopos,* or aim of religious training and devotion. Diversion of the mind from its purpose was, for Cassian, a form of *fornicatio*, a wantonness. Mary Carruthers (1998) explores Cassian's writings and documents how curiosity interfered with the practice of meditation in medieval religious communities. Among the "arts of memory" necessary to attain true piety when approaching the Word of God as imparted in the sacred texts, disciplined attention is essential. For Cassian "a mind which lacks an abiding sense of direction veers hither and yon by the hour, and by the minute is a prey to outside influences and is endlessly the prisoner of what strikes it first" (p. 82). Special faculties of the mind were developed in mnemotechnic patterns and structures, or "bins," organizing the content of scholastic study into what, in contemporary cognitive terms, are referred to as conceptual schema. For church pedagogues like Cassian the mind is naturally inclined toward an entropy of laziness and listlessness. This downward spiral of a dissipative mind is manifest in a novitiate "dilettante and a nibbler on spiritual interpretation" (p. 83). These errant novices are continually tempted by distractions, whose thoughts merge in acts of nothing less than "mental fornication." Writing before Augustine, and an important influence on the formation and rules of the Benedictine Order, Cassian chastens monks in their meditation and readings, enjoining them from a cognitive vice characterized as "an aimless, fruitless, shifting expenditure of energy" (p. 82).

Constituting nothing less than a "pact with the Devil," Blumenberg cites the charges levied against curiosity and its practitioners by church fathers who may have considered it a "compensatory extravagance that provides itself with a substitute, in the enigmas and mysteries of the world" (p. 335). This vice figures decisively within the medieval episteme also as a diminutive expression of *acedia*, the sloth and "flight from purpose" that continually threatens the *studiositas* of monastic instruction. It is a vice inscribed in the legend of the wandering monk, in a straying of the will from ecclesiastical canon, habit, and comportment as well as from sanctioned reading practices. In the scholastic communities of medieval culture, curious manuscripts are forbidden from the trivium and quadrivium. A general mistrust of subjects and domains of knowledge, including such particulars as the long-standing mistrust of astronomical curiosity, is evident in their demoted status within the scholastic curriculum.

The dangers posed by the ever-present temptations of sensuality, idleness, and aspects of an abiding *acedia* preoccupied the early church fathers. Acedia constituted a pivotal role in the lure of the deadly sins. In Reinhard Kuhn's (1976) historical profile of acedia, a forerunner to modern expressions of melancholia and ennui, the patristic accounts are fueled in large measure by their exhortation to acolytes to resist this "demon of noontime." Anchored in biblical scripture and its disciplined reading among isolated desert communities of monks, St. Paul, St. John Chrysosthomos, Stagirius, Cassian and other early Christian leaders could and did seize upon received doctrine. St. Paul, in his advice to the Thessalonians (II Thessalonians 3:2) repeats the apocryphal exhortation of Ben Sira with regard to an indolent, metaphoric servant: "Put thy servant to work, that he be not idle; for idleness teacheth much mischief" (Ecclesiasticus 33; Kuhn, p. 27).

An important recurring theme concerns questions of how knowledge is institutionally brokered and negotiated. In taking an admittedly narrow domain of cultural history, both Eurocentric and literary, and excluding the founts of crosscultural and popular cultural expression, I adapt Gellner's (1988) use of the term *clerisy*. It identifies shifting constituent technicians of the word, specialists, and appointed arbiters in "cognition, salvation, ritual." In their differing roles across historical and social formations, clerisies function in the administration of knowledge and the conservation of specific symbolic practices. In the context of the early modern period, a clerisy of church fathers—including *docti, savants,* "men of letters," clerks, and masters—acting under or against the frequently contentious authority and alliances of church doctrine and the force of sovereigns, adjudicated the rules and canon of reading practices. They fixed and reformulated the pedagogy and hermeneutics guiding novitiates and advancing acolytes in their path of spiritual progress. This clerisy, under new secular and social forces during the twelfth and thirteenth centuries, was concerned with the preservation of traditional principles in structuring the course of authorized reading within the monastic and convent communities. Their primary concern was to augment attention to the holy and sacred within the activity of textual experiences as aspects of prayer, study, and meditation. The formation of a canon consists of more than the proscription of excluded texts as heretical, blasphemous, or pagan. At the edge of the feudal age, in the sponsorship and valorization during the Enlightenment, and in the early modern period discussed in the next chapter, changing parameters of inquiry, imagination, and interpretation will appear under altered epistemological and moral points of reference; in each a clerisy most importantly champions practices for the reception and transmission of select knowledge as forms of social power. It is a clerisy's vested power to establish, contain, and enforce the rules for engagement in practices that formulate the method, form, and content for participation in its claim to truth.

The permissibly allotted domain of human inquiry, however, expanded as an unexpected historical response to the censorship by Tempier, the Bishop of Paris and the faculty of the Sorbonne. Intended as a vigorous theological response to the exponential acceptance of Aristotelian natural science by High Scholasticism, the Condemnations of 1277 led, ironically, to the pursuit of the questions of the nature of the material universe and the possibility of multiple worlds. The work of scientific speculation arose gradually over succeeding generations, in tenuous, cautious openings. Often regarded as precursors to experimental scientific inquiry, Robert Grosseteste and especially Roger Bacon emerged within a theocratic culture imbued with natural magical practices. Their sense of experimentation was applied to finite perception, mechanics and a recording of divine design. This Franciscan stated that "without experiment, it is impossible to know anything thoroughly" and through them we become "experts" (in Thatcher, 1907, p. 369). Hadot points out that though he was certainly not the inventor of the experimental method, his vast works pursued a certainty outside the confines of scholastic dialectic. In interests like his observations of the colors of the rainbow and the movements of water dripping from oars, he articulated an *experimentum* that signified experience and immediate knowledge. His "experiments," consisting essentially in practices of natural magic, a form of mechanics, and made for astonishment for divine power, were intentionally opposed to abstract and purely rational knowledge.

As commanded by Hugh of St. Victor in the *Didascalicon*, the monk and seeker are beseeched to temper and sharpen their wits against the multiple temptations for the mind to stray, and to secure for themselves the limits of what God has permitted to be knowable to humans. In the ascendance of a trained scribal class, the twelfth century marked a new integration in the formation of "textual communities." For Brian Stock (1983), the unprecedented purposes, ordering, and organization of written documentation and of the records of events occurring at this time, in both ecclesiastical and secular contexts—the origins of clerks and lawyers—constituted "one of the few universalizing forces that Western middle ages knows as a whole" (p. 18).

Hugh of St. Victor's *Didascalicon* is worthy of our attention in that it presents one of the first instructional guides for disciplined reading, study, and knowledge. It is an expression of patristic anxiety over the influence of the scribal culture's capacity, through its readership and through the quantitative increase in the production and distribution of manuscripts, to violate increasingly fragile canonical borders. The threat of "profane" classical texts remained as pressing in the time of Tertullian, Cassian, and Augustine—so did challenges to not only orthodox interpretive practice, most notably by Abelard's innovative dialectic, but also the complexly intertextual and crosslinguistic reception of classical manuscripts preserved and translated by North African and Near Eastern scholarly communities, particularly the newly available works of Plato. Censoring through exhortation toward *studiositas* needed periodic reinvigoration in order to ward off the ever-present lure of acedia and *curiositas*. Hugh of St. Victor's *Didascalicon* can be interpreted as working to insure the proper traditional balance of study and providing the guidelines for contemplation of scripture and the moral rectitude of the brotherhood. It also redefined the borders and boundaries of scholastic study as opposed to the threats to its integrity. The *Didascalicon* presents a guide to what the Abbé considered a natural unfolding of wisdom. It reestablishes both a canon and prescribes a textual community's interpretive practices in the unfolding of scriptural truths. Hugh of St. Victor praises the development in scholars of a religious or *holy curiosity*, explaining that the purpose of meditation is to foster the "ability of a keen and curious mind to explore obscurities and unfold perplexities" in the Scriptural canon (Carruthers, p. 99).

Medieval *memoria* and meditative practice composed an oppositional apparatus, countering the forces of mental and spiritual distraction. To withstand the temptations of a distracted attention, or of a will alloyed with human passions and attachments, elaborate practices were performed. In contrast to the warnings against ocular lust and wantonness, as found throughout traditional patrician discourse, cura partook its classical connotations as a form of carefulness, crafting and attending to spiritual purpose. In the arts of memory this was more commonly referred to as *sollicitudo*. Within the canonical fixing of textual authority, the *Didascalicon* is capaciously liberal in exhorting monks to "learn everything; you will see afterwards that nothing is superfluous." Hugh of St. Victor (1961) entreats the monastic community to a study as physical as it is moral or intellectual: "the field of your labor, well cultivated by your plough, will bear you a manifold harvest" (p. 139). It is an intensive regime, disciplining reading for the monastic life of journeymen of the spirit.

The recurrent image of wandering was invoked by the church fathers in shaping the life of monastic study. As both Mary Carruthers and Christian Zacher illustrate, the importance of monastic *stabilitas*, of living under the Benedictine Rule, was central. Under the scepter of such leaders as Bernard of Clairvaux, erring from disciplined knowledge was personified by the *gyrovagi*, wandering monks who partook of the lessons of partially independent scholastic institutions and masters, such as Abelard, giving rise to the first universities. Church doctrine resisted the unfolding of knowledge without a sanctioned grounding in pursuit of the holy and divine. Any of the church fathers could command their flocks to restrict their lives of study, calling upon scriptural citations such as the one from Ecclesiastes, which lament "for in much wisdom is much grief; and he that increaseth knowledge increaseth sorrow" (Ecclesiasticus 1; 18; Kuhn, p. 60). For Bernard, curiosity is personified by "the portrait of a monk who wanders with inquisitive eyes, his head cocked expectantly and his ears perked up wherever he goes" (Zacher, p. 25).

For Hugh of St. Victor the somnambulist pilgrim typifies the vagrant, restless condition of a seeker. In his text, seekers of knowledge and truth must become wanderers who are estranged from their native land. He encourages complete detachment and spiritual wandering for acolytes in which "all the world must become a foreign soil for those who want to read with perfection." The poet says: "I know not by what sweetness native soil attracts a man. And suffers not that he should ever forget. The philosopher must learn, bit by bit, to leave it" (1961, III, 19, 101). To cure the wont of wandering and the dominating power of the eyes, an outward manifestation of inner mental wandering, the monk must, as Benedict said, keep "head bent and eyes fixed on the ground" (Zacher, p. 24). This is accomplished through the disciplining of mental habits and the structuring of daily practice, through a *cura*, a caring for the sanctity of the devotional life and a policing of the mind and body in their often-deleterious passions. The mind was inherently prone to be distracted by images, color, and icons that the church fathers found increasingly present in their cloisters. In the arts of memory, the images presented by sculpture, murals, and illuminated manuscripts could entice the mind to wander and inflame with phantasms the ever-present passions of the monastic community. In the context of this iconoclasm, Bernard exhorts "what is this monstrosity doing in the cloister, in the presence of the brothers while they are reading" (Ginzburg, 1976, p. 32).

The injunctions of the church fathers and doctrinal faith held sway for nearly a millennium. St. Paul's exhortation, *"noli altum sapere, sed time,"* contains a complex layer of interpretations and translations guiding moral and intellectual behavior. Carlo Ginzburg surveys the impact of St. Paul's call to "be not high-minded, but fear," in Jerome's literal Latin Vulgate translation of the original Greek. In the early years of the Church the moral resonance of "sapere" was gradually transformed to connote the zeal of epistemological ambition. For Ginzburg, a formative misunderstanding of this passage came to comprise a significant element in early Christian literate culture. He indicates that by the fourth century this passage was recoded, with Ambrosius, for instance, proclaiming the phrase to mean that "it is better to fear things than to know them." This switch in semantic value assisted the general development of a canonical corpus of readings for monastic culture and a constriction of the realm of permissible inquiry. This transition signifies for Ginzburg that "St. Paul's condemnation of moral pride became a warning against intellectual curiosity." A singularly influential reading, the slippage from moral to intellectual sanctioning held its power for "centuries and centuries, by lay as well as ecclesiastical writers, as the standard authority against any attempt to overcome the boundaries of human intellect"(Ginzburg, p. 37). Ginzburg cites Nicolo Malermi, among the first Italian translators of the Bible, declaiming in the fifteenth century *"non volere sapere le chose alte,"*—"do not seek to know high things." This was to remain a doctrinal premise until Erasmus's clarification that the intent of St. Paul was to direct human activities away from moral vices, not intellectual pursuits.

Among the enduring metaphors through which curiosity has recurrently been debased are those that present it as a form of appetite or, in Nicholas Barbon's phrase, a "wanting of the mind" (in Berry, 1994, p. 114). The generally dismissive regard, which likens curiosity to the organic needs and drives of the body, bears the mark of a patrician legacy, most notably of Augustine's injunctions against the power of this "lust of the eyes." This is the same church father who confesses to his own acquisitive appetite as a youth in the story he relates about stealing pears.

The appetitive function in particular, however, is particularly resonant. It is a trope in disciplining the literate monastic orders, one in which the iconography of Luke digesting scripture would be well known—an image that may have worked not only to forestall curiosity's tempting diversions, including the usually forbidden "pagan" and "profane" literature of antiquity, but also to consummate canonical injunctions by incorporating, digesting, and becoming one with the Word of God. It is, perhaps, a miniature of the *imitatio dei*, a transubstantial act of taking in and making sacred text the material body itself, as flesh: the passion of the curious aspirant stands tensely between holy and profane texts, consumed in an act mirroring the Passion itself.

Notes

1. In Snell (1953), *theoros* implied a minimal level of conceptual order and schematization. Other cognates identified very specific kinds of vision with sight being always determined by the type of object perceived and sentiments regularly accompanying them. Snell provides one in particular, *paptainein*, a mode of looking about inquisitively, carefully or with fear, suggesting the perceptual and physically embodied rootedness of theoretical inquiry.
2. As the intellectual faculty of the psyche, *nous* is also semantically related to *gegnskein*, to recognize (as in identification of a man). Snell (1953) states that *noein* refers particularly to situations—to acquire a clear image of something, stressing the mind as a recipient of clear images organ of clear images, the mental eye. It connoted the function, but not necessarily the organ of perception and the power of intelligence.
3. Sense perception, including the emotional currents of *thymos*, was also regarded as conscious perception, a form of thought and mental activity that parallels the medieval scholastic concept of *haeccitas*.
4. It is not clear how Lloyd and Sivin (2002) distinguish this manifold of structures of knowledge from earlier formulations, as those made by the French sociological school of "mentalities." These structures are economic and epistemic premises of research programs as well as their methodological protocols and utility in their historical context. Lloyd and Sivin may be countering what they perceive as a rather abstract and continental, mostly Foucauldian-inspired critical historical writing about "epistemic regimes."
5. This term is the equivalent to *periergia*, overcarefulness, superfluity, investigation of curious matters. Blumenberg (1985).
6. Katzenellenbogen (1939) describes the *Psychomachia's* significance in defining the stylistic criteria of medieval iconography.

Chapter 2
A Taming of the Passions

The purpose of this chapter is a partial mapping of theoretical curiosity during early modernity. I continue the genealogical approach of the first chapter in gauging significant aspects of its participation within the period's formative and unstable conceptual territories. A genealogical approach is helpful, I believe, in my attempt to demystify the peculiar status of the forays of theoretical curiosity in contemporary culture; its uncertain place resonating in the hopes, questions, and anxieties about the uses, effects, and ethical limits to technological research.

As a starting point for appreciating the effects of the dramatic revaluation of theoretical curiosity in early modernity, I will profile here some pivotal historical and epistemological changes occurring across early modernist and Enlightenment intellectual and cultural terrains. In subsequent chapters I offer profiles of the ways in which some distinctive contemporary cultural and technical practices, in particular pedagogy, genetics, and artificial life, become issues for general, public, and critical attention. I favor a process of historicizing epistemology in the manner that Hans-Jorg Rhineberger (2010) has advanced in his extensive critical ethnographic work in laboratory research in molecular biology and the "life sciences." I am especially interested in emphasizing how theoretical curiosity assists the critical interrogation of existing knowledge and social relations. I maintain that theoretical practice is never fully removed from the social conditions and communities of practice from which it emerges. There is no single agency or factor that determines if and how innovative insights and developments might be put to use and whether their social effects are destructive or ameliorative. I wish to stress the interested nature of theoretical work and its shaping power in the making of real and potential ways of knowing and becoming.

My critical theoretical interest is directed to thinking how theoretical practices interconnect with the nature of everyday life, its rhythms, order, and potential for transformation. I advocate for a socially engaged *critical curiosity* that plays an often-hidden role in the creation of concepts and practices. A critical, performative orientation is intended to "cut off the King's head" in the long sovereignty of a sterile and ahistorical epistemology. It is a guiding thread in this project's navigation of the unchartable continua of experiences, paths of inquiry, and thought experiments within specific communities of contemporary cultural practice.

M. Zuss, *The Practice of Theoretical Curiosity*, Explorations of Educational Purpose 20, DOI 10.1007/978-94-007-2117-3_2, © Springer Science+Business Media B.V. 2012

The seventeenth and eighteenth centuries manifested cultural and epistemic shifts, cresting with the emergence of Enlightenment practices. One of the many currents running through modernity's pivotal historical and conceptual transformations was the concerted social project directed toward a "taming of the passions." This domesticated "affection," derided in dominant classical and theological evaluations of the place of the senses, gained stature. It performed as an activating agent in the processes of newly presiding empiricist modes of inquiry.

The unquenchable thirst or appetite for knowledge long under prohibition, suspicion, and constraint takes on what I consider new material practices or conceptual personae. These are put into play in the renaissance with early modernism's sponsorship of empirical inquiry and artisanal practices. The release of Pandora's amphora is read especially as a quarrel between the Olympian gods and humans after Mercury grants Prometheus the gift of fire and the arts of material culture. Zeus, seeking his punishment, has Vulcan fashion both Pandora and her notoriously alluring vessel. While Prometheus refuses Pandora's charms as well as her infamous "box," his brother Epimetheus accepts them. Prometheus, who according to Paulo Rossi (1962), was traditionally the icon representing the rebel creator of a culture distinctively human through the imparting of the potent knowledges necessary to master nature, is gradually transformed by the middle ages to an inventor of arts and the cultural resources making civil life possible. Rossi asserts that the increasing importance of Prometheus to early modernist practices expressed emergent cultural and moral changes that foregrounded humanity in the scheme of the world. As makers of possible worlds, human knowledge is a functional power that participates in the chains of causality giving shape to the existing material world engendered by the acts of conscious and willful gods. Pandora's vessel, a prototype of the age's distinctive cabinets of curiosity itself, is also an icon, I believe, of early modernism's vigorous embrace of *techne*, the means of material production.

Pierre Hadot (2006) describes two originating attitudes that will be useful for my discussion of the uses and pursuits of theoretical curiosity. These two attitudes or dispositions, which he terms the Orphic and the Promethean, are collective motivations. In his history of the "unveiling" of Isis, the intentional disclosing of the "secrets" of nature that are the theme of Hadot's dazzling essay, the Orphic attitude or persona is one that would open secrets through a respect in wonderment and mystery (p. 95). Its inquiries, as in Aristotle and the Presocratic physicians, purport to be disinterested. The Miletians in particular, discussed in the first chapter, evince this attitude and relation. In large measure their inquiries and "histories" seek after the patterns of universal processes. In their common, though not entirely speculative endeavors, Thales, Anaximander, Anaximenes, and Heraclitus marvelled at a *physis*, the generative first principles of things in the natural and material world they wished to decipher and disclose.

In my tracing of some of its contemporary personae in Chapters 6 and 7, of contemporary technoscientific inquiries, the reader will note that the family quarrels between the gods and humanity remain unsettled today. There is an epistemological and ethical chain of relations that leads to a continual dispersing of Pandora's potent calamities on the world. Among the many that could illustrate this effect, is the deserved attribution of a Promethean will to the twentieth-century physicist Robert Oppenheimer and the Trinity Project, which brought the primal blazing of nuclear fire to realization. Through the gaze of a theoretical curiosity that was, ultimately, neither limited or averted, knowledge and power fused in the formative power of fire in the myth of Pandora. For Rossi

> the fire stolen by Prometheus is the origin of industry—of the mechanical arts and sciences—through which man can modify his situation of original nakedness and defenselessness to make himself in some way the Lord of all created things (p. 183).

This theme will riddle through modern debates in Chapters 6 and 7 on the question of limits to theoretical curiosity in the technosciences.

Plato's *Timaeus* is an Orphic text in that it attempts to model itself on the design of the formation of the world. A poetic and contemplative attitude can also be found in the work of the Epicureans and Stoics. In the context of my inquiry into the desires *of* and *for* theoretical research, the classical Orphic persona accentuates an ethical stance (from my anachronistic view) as well as the arts of enchantment. In contrast to this poetic and "philosophical physics" but never, I claim, in absolute opposition, the Promethean attitude is premised on the ruses and tricks that will "tear nature's secrets away from her, for utilitarian ends." It employs any of the *techne* and procedures for an instrumental physics. It is motivated by a will to objectify nature. In early modernity it is actualized in the will to power of a colonial order's productive harnessing of the passions of the mind. Hadot draws the difference in what I consider cultures of inquiry, remarking that Orpheus

> penetrates the secrets of nature not through violence but through melody, rhythm and harmony. Whereas the Promethean attitude is inspired by audacity, boundless curiosity, the will to power, and the search for utility, the Orphic attitude, by contrast, is inspired by respect in the face of mystery and disinterestness (p. 95).

The Orphic and Promethean persona are fraternal twins. They cohabit the same research program and processes of discovery. Thinking through them is helpful in delimiting the extremes and limits of forms of rationality and irrationality alike in the adventures of inquiry and discovery across very different historical and epistemic regimes. They converge ambiguously, vacillating between "pure" and "disinterested" speculative research, or imminent applicability. This twin polarity exists across many occasions for research, from classical and Presocratic astronomy's uses in warfare, into contemporary nanoscale instruments that can manage health as well as surveillance of daily life. It is a tension between what we now regard as a speculative research and development orientation, "pure science," and the pursuit of those Stephen Shapin (2008) documents as the "angels" and entrepreneurs who are aggressively and competitively courted for start-up technoscientific projects.

These conceptual persona offer germinal figures for the methods, processes, and structures in which research is conducted. These dispositions can be institutional or act as counterdiscourses to the prevailing practice and cultural tradition. In their Promethean aspect they conduct emergent technical developments toward specific aims. In early modernity this persona becomes dominant in engineering the resources necessary to sustain and build new forms of research for mastery over nature. By Francis Bacon's time, the myth of Prometheus is a figure for the inauguration of a decidedly experimental science. As Lenoble succinctly phrases it, "in the seventeenth century Prometheus becomes God's lieutenant" (in Hadot, p. 96).

One precursor, Nicholas of Cusa, personifies the turn from medieval speculation to an engaged theoretical curiosity. At the cusp of this transition Cusa (1401–1464) could be thought as a pioneer of curiosity, the "first modern thinker," marking the advent of genuine intellectual and theoretical pursuits. For Cassirer (2010) he was the first to venture questions that were either forbidden or kept unspoken in the medieval context. This was made evident in his "asking not about but about the possibility of knowledge about God" (p. 10). His distance from our purportedly secular age makes it difficult to sense how radical a question like this is posed for his time. Its mere posing expressed a not entirely "holy curiosity" that could have caused him censure. He pursued careful inquiry into the "absolutely greatest" and the "absolutely smallest" aspects of nature, knowledge, and experience of divinity. His concern for the limits of knowledge led to his articulation of a mathematical model for contemplating infinity. Breaking with scholastic doctrine and dialectic, his *On Learned Ignorance* sought a unity of oppositions, one which could be approached through the "self-movement" of the mind in the perpetual processes of thought.

Cusa is important in tracing the legacy of research in his advocacy of an intellectual vision. His is a sense of a purely qualitative, continuous, and unlimited process of thought. Thinking participates in the infinite and cannot be reduced to quantitative pursuits. In this he is a precursor of all attempts to elude the snares of epistemological closure. He proposed oppositions that coincide, in a continuum between the Absolute Smallest and the Absolute Greatest or Maximum. The Maximum is purely qualitative, the absolute ground of being and thinking. The place of the sensible is revived here. In these distinctions he prepared the ground for methodical measuring techniques of early scientific projects. The conditioned, derivative world of sense perception and earthly experience is reinvigorated as the starting point to launch all research. The sensible exists only within the limitlessness of being. For Cusa all knowledge remains conjectural, formed in human attempts and their mere probabilities. His method breaks with classical and scholastic tradition. He assigns an initial relativism in positing points of reference in the mathematical determination of finite relations of space. Infinity comes to bear upon the lived experience and the limitless search for knowledge. All visible things are images of the movement of the invisible. The indeterminate nature of empirical things stands in opposition to a supersensible infinite and yet functions in what he formulated as a "coincidence" of opposites rather than the formal and mutual contradiction of scholastic dialectics.

The infinite begins to participate in the adventures of the emerging era. It will come to play an essential role in the relations established in the increasingly abstract calculations and concern for measurement and proportion central to the nascent sciences, theories of knowledge, and art. It will come to inform the concepts of coordinates in Descartes and Fermat, the theories and observations of movement and motion of Galileo, the calculations of planetary orbits by Kepler as well as the working premises regarding the creation and making of forms for Leonardo in the aesthetic sensibility of the Renaissance.

Cassirer (2010) remarks on the formative change of thinking made by Cusa:

> The mind can come to know itself and to measure its own powers only by devoting itself to the world. Even sensible nature and sense-knowledge are no longer merely base things, because, in fact, they provide the first impulse and stimulus for intellectual activity. The mind is the living illustration of eternal and infinite wisdom; but until it is stimulated to movement by that admiration, which arises from contemplation of the sensible, it is, so to speak, asleep within us. This movement, which begins and end in the mind itself, must pass through the world of the senses. The mind always assimilates itself to the sensible world; mind becomes sight when presented with color, hearing when presented with sounds. The descent into the world of perception is no longer considered a decadence, a kind of sinful fall of knowledge; instead, it accomplishes the ascent of the sense-world (p. 44).

Like Paul to the Corinthians, Cusa accepts the premise that we cannot see except in a mirror or in an enigma. We must surpass the veil of images covering the face of the certain and immutable. This is only possible through the vehicle of symbols and sense images in order to go "beyond all faces" (p. 33). We face the world when we ascertain the infinite in the finite. This notion reprises the ancient image of the human as a microcosm of the universe. The limitless and unbounded, Anaximander's apeiron, is ingrained in the skeins and textures of the material world. As Cassirer emphasizes, the finite world as a reflected splendor of the divine or infinite is a commonplace premise in medieval and hermetic practice. Cusa refigures this image, boldly endeavoring to begin the investigation of nature. Isis never reveals herself and the book of nature, written with the signature of God, must be researched. Holy curiosity is placed in the play of mirrors. God's work is placed under scrutiny. The divine must be detected. Cassirer claims that with Cusa we are "on threshold of open field of objective science—neither subjective feeling nor mystical sentiment suffice to understand the meaning of the book of nature. Rather it must be investigated, it must be deciphered word for word, letter by letter" (p. 53).

Pertinent to the enduring questions of perception and knowledge, Cusa also advanced a geometric model of the gaze. He distinguished sight from the gaze, the former composed of multiplicity of perspectives, mobility, and changing points, the latter structured in his geometric diagram as the unity of all possible spatial relations and points of view. This concern with the gaze and the visuality will recur in considering technology and the society of the "spectacle" in later chapters. His model is a "mathematical mirror" of the world, delineated nothing less than Cusa's contribution to an emerging specular "science of seeing" that Leonardo would later bring to fruition. Along with the researches of Alberti and Pierro della Francesca, his specular mirror reflected the fifteenth-century interest in the relations between mathematics and a theory of representation in art. As Michel de Certeau (1987) interprets these pursuits

> mathematics is not only a condition of certainty in knowledge, but the methodical way of decoding or organizing scientific "observations." A school for rigor, it is at the same time a science of seeing (a hermeneutics of figures) and an "architectural science of reasoned and demonstrative experiments. In the new scientific order, which is essentially visual, it takes the place that logic held in medieval science, which was essentially linguistic (p. 14).

Speculation is inherently limited. Despite the illusion of its open-ended range across all or any field of inquiry, speculative thought remains distanced from experience and material relations. For Leonardo, the artist fulfills what speculative thought can only keep at a distance. Speculation can indicate and refer, setting formal parameters and conceptual coordinates. It is exemplified in the logic of ideal relations, as in the geometry of the ancients which Husserl and Merleau-Ponty recall as increasingly abstract practices derived from actual work and connections to the physical earth, for instance in the building of dams on the Nile. The artist bears the ideal image and potential into material substance. This requires the processes by which human work, as techne, gives the world form. Cassirer (2010) provides an astute appreciation of the priority of the artist and scientist in Renaissance practice claiming that "man can only be certain that sense world has form and shape if he continually gives it form. Ultimately, the beauty of the sensible world does not derive from itself; rather, it is founded in the fact that it becomes, in a sense, the medium through which the free creative force of man acts and becomes conscious of itself" (p. 67). It is an era in which the Promethean, even Faustian persona, predominates. As a regime of practices, theoretical forays are released from constraint: "The proof of the mind's specific perfection consists in its refusal to stand still at any attained goal and in its constant questioning and striving beyond the goal" (p. 69).

The visual remains the pivot in a new representational order initiated by Alberti and Perelin. The visual imagination shifts its coordinates to elaborate proportion and the depths of a perspectival view with its calibrated vanishing points. Proportion, Cassirer points out, becomes a principle for aesthetics, no longer restricted to mathematics and speculative logic and dialectic. Drawing from Cusa's claim that all knowledge can be defined as measurement, the ideal of proportion incorporates the potential afforded by the arts of measurement and proportion. It becomes a Renaissance paradigm, in which the intellectual and the technical

> of the period converge in the concept of The idea of measurement becomes the connecting link, joining the natural scientist to the artist who creates a second "nature"... proportion is not only the mother of knowledge; it is the "mother and queen of art". Thus the speculative-philosophical, the technical-mathematical, and the artistic tendencies proportion (p. 51).

The ocular engages, seizing proportional relations, connections, and forms inherent to things that can be made actual. This allows for the unconstrained use and expression of the creative forces, technical and manual knowledge of artists and scientists. The Promethean persona is present here working the world into form, reflecting the face of Faust in the eye of the new discoverers, an "eye, a sense organ," which "is neither satiated nor limited by anything visible, for the eye can never have too much of seeing; likewise, intellectual vision is never satisfied with a view of the truth" (p. 69). The ocular engages, seizing proportional relations, connections, and forms considered as timelessly inherent to things.

The ancient Pythagorean fourfold division of mathematics included geometry, arithmetic, or number theory, astronomy, and music. Revived in the medieval quadrivium, it generated major resonances in the adventures of theory in social life. Astronomy, significantly contested as a model of the relation between worlds and human's place in the order of things, was considered a form of applied geometry, and music as applied arithmetic. Mathematics, for the Neoplatonists of early modernity, was primary for the gradual expression of methodologies and models that would later form a core for the natural sciences. This was particularly manifest in the reinterpretation of nature in an evolving mechanistic model or paradigm of the physical universe. In Marie Boas's (1962) estimation, mathematics, in its Neoplatonist rehabilitation "included in its competence the greater part of what the 17th century was to call natural philosophy" (p. 198). This mathematical model will be considered more in both this and subsequent chapters. Gadamer (1998) comments on the rise of the methodological focus of emergent science:

> when the new science took the path of methodical rigor, it ignited a true explosion that burst apart not only the Middle Age's geocentric view of the world, but also its theocentric one. In the seventeenth century, after Galileo, the mathematical construction of idealized relationships of motion was elevated into the method of knowing reality. It succeeded in constructing classical mechanics, which in the end, thanks to Newton's combining it with celestial mechanics, ushered in a new sense of the world that also changed the ideal of the theoretical life: knowledge became research (pp. 22–23).

Paradigmatic of the province of one strand of rationalist and deductive reasoning, mathematics was at the same time intensely employed in all aspects of practical life. The vaunted rationality of Euclidean geometry eclipsed its more Archimedean-inspired rival in a world in which seafaring and the calculation of longitudes were vital. Mathematicians like Pedro Nunez delved into methods for making navigation precise. In his *On the Art of Sailing*, Nunez developed the notion that "since the meridians on a globe converge, a true sea chart should not have its meridians everywhere equally spaced" (p. 205). He fabricated a quadrant allowing sailors to discern the number of leagues in a degree along each parallel. Jacques Bresson was an inventor of instrument for navigation, timekeeping, and astronomy. The polymath and mystic John Dee put mathematical knowledge into service, working as an *intelligencer*, in contact with Fine, Fernel, Mercator, and Hariot, among others, in developing new navigational and cartographic models and methods. In the formation of early engineering techniques, mathematics clearly augmented the powers of production in the design and implementation of pumps, lathes to cut cylinders and cones, cranes, bridges, war machinery, and pneumatic and hydraulic devices. These inventions, discoveries, and applications represented a convergence of mathematics and the work of "practical men" in the expansion of the new field of mathematical physics. For Bresson, the "contemplation of the proportions of numbers, points, and measure of artificial things is useless unless related to action, so that it follows that mechanics is the fruit of geometry, and consequently, its goals" (p. 211). Geometry maintained its status as the most advanced branch of the new mathematical investigations as most significant works in late sixteenth century were concerned with the application of geometry to optics, astronomy, and mechanics.

In John Dee's "*zographie*" mathematics was applied to art. After the problems of rendering three dimensions on two-dimensional canvas planes were resolved by Perelin, in his *On Artificial Perspective*, painters were "filled with curiosity to know why the tricks of the trade worked" (p. 200). The empirical application of mathematical knowledge to the arts was also explored by Durer and Leonardo, for whom geometry was the basis for painting as well as the crafts of building. Both Durer and Leonardo pursued ways of making these available as public funds of knowledge.

This era evidenced an avid interest in the relation between theory and practical techne in the dissemination of the new knowledges. Increasingly, a proliferation in the teaching of science flourished, allowing vernacular access for a general public and operating outside the restricted canons of a classical pedagogy in the medieval scholastic universities. A nascent, experimental science was made to manifest its usefulness. The new scientific education was not entirely formalized and institutional. It was a decentralized pedagogy, formed outside of the early universities in family arrangements like John Dee's at Mortlake, or Tycho Brahe's science center at Uraniborg with its hosts of aspirants and assistants, and in what Shakespeare dubbed Walter Raleigh's "School of Night." Uraniborg and the *Academia dei Lincei*, which Galileo attended, were perhaps the most productive cauldrons for the spirit of the new scientific curiosity in its applications to empirical method and discovery. The Academy of the Lynxes included Stelluti who is associated with publication of the first microscopial figures and where the name of telescope was first applied. For the *Pleiade* group, whose inquiries were primarily literary, a special interest in music was pursued, particularly for mathematics and physical acoustics. The aesthetic and scientific aspects of music of concern for the Pleiade group was apparent in Mersenne whose *Harmonie Univerelle* attempted a cataloguing of all known forms of instrumentation and the physics of their sound production, scales, patterns, and harmonies. He exemplified the era's "*intelligencer*," a man who made it his business to know and correspond with the leading scientists and philosophers, including Descartes, of the day. Nurenberg became a center of astrological instrument making and printing and in Paris, Dee gave informal lectures in the new sciences, while Fine taught at the College Royale, and acquired professorships in mathematics and medicine. In England, Gresham College was the center for the new curiosity in scientific astronomy and geometry.

In spanning theory, speculation and practical application through geometry, engineering, architecture, and the mechanical arts of early modernity mathematically informed knowledge functioned as a generative epistemological model of the world. It bridged the arts of techne and epistemological inquiry. To a great extent, there was a convergence between aspects of rational, speculative inquiry and empiricism. They worked in increasing alliances that helped found the self-consciousness of modernity. Their combining powers were actualized in the era's material productivity and the deployments of its technological models, tools, and instruments.

Francis Bacon, who claimed "all learning to be his province" embarked on nothing less than a reorganization of science based on empirical observation inductive method. The "new learning" made science the key to truth, with empiricism as its basis. The application of science to the useful arts would improve the material well-being of mankind. Arguing for a "luciferous" as opposed to a "lucriferous" knowledge, Boas claims that "in many ways Bacon was the real progenitor of the 18th century enlightenment" (p. 250). For Bacon

> men have entered into a desire of learning and knowledge, sometimes upon a natural curiosity and inquisitive appetite; sometimes to entertain their minds with variety and delight, sometimes for ornaments and reputation: and sometimes to enable them to victory of wit and contradiction; and most times for lucre and professional and seldom sincerely to give a true account of their gift of reason, to the benefit of men (p. 249).

By Bacon's era, a new willingness to detect the immutable, mechanical laws of the universe was sanctioned. New and emphatically empirical study, recording, and observation were granted permission by revised theological, political, and economic principles and interests. The changed doctrines, which can be dated to the late Middle Ages, and to the Condemnations of 1277 in particular, allowed for an unprecedented accommodation between theological doctrine and inquiries into the material world. Eluding its association with dubious and prohibited realms, curiosity tenuously ascended in stature in its ascent and alignment with intelligent, rational inquiry.

The new valorization of curiosity appears also in the emblem books of the seventeenth century. The figures of Icarus, Ulysses, and Prometheus in particular are given new countenances. The title page of Bacon's *Great Renewal* bears the motto, "Many will pass through and knowledge will be increased" surmounting the image of Odysseus's ship passing beyond the limits represented by the Pillars of Hercules. "Plus ultra" became an article in the new faith of empirical science's careful transgressions into epistemological and moral territory, formerly the domain of theology and its clerics. For Ginzburg (1976), "during the seventeenth century Icarus and Prometheus became symbols of a powerful intellectual drive towards discovery. In a dramatic shift of values, 'boldness', 'curiosity', and 'intellectual pride' were now seen as virtues" (p. 38). In one illustration, Icarus appears as a winged young man floating gently in the air, encaptioned with "*nil inquire anisum,*" to dare everything. Ginzburg traces this caption to the historical exploration of the new world by Columbus. The value of intellectual risk reprised the new valences attached to interests and luxury in the pursuit of an increasingly unfettered quest for trade and commerce in the building of empire.

A shift in the basic coordinates of visibility and perspective was initiated by Perelin and his disciples. It is a key feature expressive of a new worldview and paradigm of knowledge. It is a shifting of the cardinal points designating the known and the unknown, the visible and the invisible in speculations as often based on new instruments, mathematical and mechanical models, as on enduring theological currents. As Edward Casey (1997) demonstrates, the nature of place, of *topos*, originating in the closed and contained world and cosmos of Aristotle's *Physics,* dramatically changed with a theological coup in the thirteenth century. The Condemnations of 1277, which disposed of the limits imposed by Aristotelian natural philosophy on the omnipotence of God, permitted a displacement of place as a primary element in consideration of the nature of the physical universe in favor of the primacy of space as an expression and manifestation of the divine mapmaker of the cosmos. If it is at all useful to consider Duhem's claim that the Condemnations were the "birthdate of modern science," it is only because they opened up previously closed conceptual schema for multidimensional perspectives that would rapidly emerge in the succeeding centuries. They mark a pivotal transition point for what Casey considers a progressive community of scholarship and speculation that will encourage the "thought experiments" of Bradwine, Crescas, Cusa, and Bruno, arising from the new conceptual arena regarding the nature of material existence and its manifestations. For Casey, these thought experiments proved to be basis for a reorientation of the core terms by which human inquiry into nature were to be derived, "engendering the conceptual ventures that laid down the foundations of modern physics, above all its commitment to the infinity of the physical universe" (p. 107).

The new concern for spatiality and perspective complemented and derived the emerging European states and their fledgling empires. Place, conceived as a perimeter, enclosed and limited, in the fixed, finite cosmology of Aristotle, was to become

> an adventurous avenue toward infinite space opened up decisively after the thirteenth century in the West. The closely confining circuit of place-as-perimeter dissolved and the vista of a New World of Space began to captivate ablest minds of the succeeding periods. It hardly seems accidental that the great Age of Discovery in the fifteenth and sixteenth centuries-an age that set out expressly to explore a terra incognita of interconnected places within the larger space of the earth itself as we as the still larger space of the heavens – immediately followed upon and the bold speculations of the philosophers and theologians in the 13th/14th centuries. From an entirely imagined and divine status that was fully gained by A.D. 1400, such places became actual in the form an earth and a sky that lay ready for discovery and possession not only by thought and faith but also by arms and men (Casey, 1997, p. 115).

As epistemological curiosity was invested, made virtuous, it gained a patina that now signified the exploratory spirit of prized secular adventures. It worked to goad a surpassing of all existing physical and epistemological boundaries. Serving the emergent practices of the seventeenth century, it is a yeoman whose fealty allows for invention, trespassing into formerly forbidden or unknown realms. These included mathematical and symbolic notations, the collection of oddities and artifacts. *Curiositas* is domesticated, tamed, and recuperable in the ornaments it is granted by the new sciences' structured sobriety and methods.

In the wake of Galileo's overturning of heliocentrism, the distinction between realms of possible knowledge and inquiry become increasingly explicit and stipulated. For the Italian Jesuit Cardinal Sforza Pallavicino, the conventional analogy between "arcana naturae and arcana imperii," the secrets of Nature and the secrets of political power are redrawn. Nature was predictable within a cosmological system of predetermined mechanical principles documentable by finite human minds, but "to predict the behavior of kings and princes was sheer temerity—as it would be to predict God's inscrutable will" (Ginzburg, p. 37). It is a contrast predicated on an implicit injunction to keep the "common people" from intervening in worldly affairs.

Reading Republics

During the formative years of the Enlightenment, investments of capital were required for an increasingly complex division of labor requiring specialists for new literacies who could compile, edit, amend, and revise the various editions of these often-multivolume ventures. In addition to such popular encyclopedias as Pivati's *Nuovo Dizionario Scientifico* and the *Lexicon* by Zedler, multiple editions of Chambers's *Cyclopedia* and the *Britannica* were extremely successful in dispensing collected editions of knowledge to an expanding public. Most influential among them, however, was Diderot's *Encyclopedie*, which by the eve of the French Revolution had been produced in several cities in varied editions amounting to over 25,000 copies. The newly codified editions of the seventeenth and eighteenth centuries differed from their predecessors by sanctioning an explicitly useful and progressive mode of inquiry. Where their medieval counterparts were construed as models of the cosmos in miniature, intimations of the supersensible and placements of worlds within worlds, the new forms of curiosity were crafted, suited to the new systematic projects of the era. These projects engaged a mobilization of texts in the interest of a new distribution and transmission of textual literacy within the circuits of knowledge and power made necessary and desirable to a nascent colonial order.

The appetite for the new consumption of knowledge coexisted and thrived with institutional forms that aided, benefited, and promoted the production of consumable knowledge. The eighteenth century witnessed the establishment of academies for the arts in London, Madrid, and Venice. Academies and proto-universities ensconced the new *virtuosi*, savants and clerisy in Berlin, adding to power and influence emanating from already-existing centers in Paris, Bologna, and Florence. A significant semantic shift occurs in the value accorded to research and its cognates. Research, *rechereche,* or *ricerca* was more commonly used as a plural term coming more widely into use during this period in association especially with investigation and experiment. Burke (2000) comments that "this cluster of terms suggests an increasing awareness in some circles of the need for searches for knowledge to be systematic, professional, useful and co-operative" (p. 46).

Theoretical and epistemological curiosity is harnessed, their formerly vilified association with the temptations lurking among the idle, now operating for the production of empirical knowledge. The increasing separation and uneasy dominance of intellectual labor over artisanal techne and manual labor paralleled new demarcations for the division of knowledge and practice. The empiricist revaluation of knowledge ushers in an appreciation for and control over the value of practical knowledge of the mechanical arts. The natural sciences, with mathematical models as paradigms of rational order, quantification, and measure, develop with a will to regulate and guide the speculative drifting of idle and pure curiosities. This turn from curiosity to research, occurring around 1700, is evidenced in Leibniz's advocacy of establishing an Academy of Science in Berlin, writing that its essential purpose would be the development of research in "contrast to mere curiosity" (Burke, p. 46).

Hirschman (1977) offers insight into how, during the eighteenth century "the passions were rehabilitated as the essence of life and as a potentially creative force"

(p. 47). This is foregrounded by Hobbes's positioning of curiosity at the core of human existence. For Hobbes (1968), curiosity is significant, insofar as it is a passion eminently employable in the formulation of a social contract. Existing as a seminal force prior to its social or political organization, he defines curiosity in the *Leviathan* as:

> Desire to know why and how, curiosity; such as in no living creature but Man, so that Man is distinguished not only by his Reason, but also by this *singular Passion* from other Animals: in whom the appetite for food, and other pleasures of sense, by praedominace, take away the care of knowing causes; which is a Lust of the mind, that by a perseverance of delight in the continuall and indefatigable generation of knowledge exceedeth the short vehemence of any carnall Pleasure (p. 124).

This singular passion for knowledge is always entangled with other human appetites and aspirations. A passion can never sustain the illusion of a purity; immersed within a social matrix, they mix with prevailing forms of power and privilege. A rare convergence of the passion of curiosity and its more august double, wonder, have been remarked upon by Daston and Park (1998) in their luminously documented history of emergent natural science. Philip Fisher (1998) treats the dalliance of these passions in more aesthetic and formal terms. Fisher, retracing the long pursuit, from Hesiod to Descartes, of the mathematics of the light of the rainbow (an interest in the practice of optics in early modernity), regards the convergence of curiosity and wonder as a natural process of thought. In his *Passions of the Soul*, Descartes (1989) states that

> whenever the first encounter with an object surprises us, and we judge it to be new or very different from what we knew before or even what we had supposed it to be, we are caused to wonder at it and are astonished at it. And since this can occur before we know whether this object is useful to us or not, it seems to me that wonder is the first of all the passions (p. 52).

For Fisher, adapting Descartes's commitment to a method of rigorous method of rational inquiry, in which "the pleasure and interest we take in the rare and unusual is part of a purely intellectual curiosity that seeks to make sense of whatever is new within experience by means of understanding" (p. 48). The spectrum of response to the world's phenomena, wonder, recognized as "a technique of curiosity" (p. 49), blends in an epistemological process progressing toward causal explanation. In Descartes's formulation, "wonder has no opposite," devolving into the habits and excess of the "blindly curious, for those who investigate rarities only to wonder at them and not to understand them" (p. 61). For Spinoza (1992), it represents a stasis of the mind, a sudden "surprise" to the soul, interrupting thought's connections. An impulse to make sense of the wondrous and astonishment motivates what becomes a rehabilitation and cultivation of the extraordinary and the unexpected. In this "passage from wonder to thinking" and in the "drive within wonder toward curiosity," Fisher detects the historical search for explanatory methods, whether religious or secular, that retraces the modes through which to witness a "kidnapping the energy of wonder" in the formulation of discursive, rational thought (pp. 40–41). Daston and Park discern a patterned synthesis of these passions in the seventeenth century, in a brief liaison short-circuited by the inauguration of the mechanist and empiricist New Philosophy.

Passions, Affects, and Social Space

Once current coursing within early modernity's body politic is the discourse of interest. In many enterprises the forces giving shape to interest parallel those that coalesce to domesticate curiosity. A "general," common or paramount "public" interest turns the new Jacquard looms and lathes running the forges of manufacture for an industrial and colonial order. A social cohesion of interests is increasingly delineated by the productive and epistemological machinery that compel a "taming of the passions." Wealth, luxury, and exhibitions of ambitious accumulation, including overt expressions of avarice, are revalued in a social imaginary that no longer strictly condemns their pursuit. It is a transformed ethos, one in which individual "interests" and social cohesion are opposed. Individual passions are harnessed in order to yield more productive, profitable social relations. In their earliest formulations in the sixteenth century, interests "comprised the totality of human aspirations—but denoted an element of reflection and calculation with respect to the manner in which these aspirations were to be pursued" (Hirschman, p. 32). This is a domain where the shadows of the seven deadly sins still stalk Christian conscience. Preceded by the practice of usury, interest was the "prince of the prince," its economic register increasingly emphasized by articulations of a machinery that at once tamed the passions and sets generated social and political contracts.

Evidence of the shifting meaning, value, and uses of the passions is demonstrated by the changes attached to the status of luxury. Christopher Berry (1994) outlines a gradual, though surprisingly rapid, "demoralization" of this traditionally vilified vice. The discourses of trade, in debates over balancing imports and exports and mercantilism, by Nicholas Barbon and Thomas Mun in the seventeenth century and by Mandeville, Hume, and Smith in the eighteenth, variously reinvent the semantic value of luxury, avarice, and accumulation. It is a revaluation of gain in the service of human nature, and one in which the simile between an individual's expenditure and that of a nation can be aptly applied. Central to the processes that launched this new valorization of luxury is what Berry regards as a generalized "re-evaluation of mundane life" (p. 114).

It is this context that the persona of the Collector dominates. The accumulation of objects of all kinds, particularly of useless exotica, including incidental material and discards from the age of exploration and conquest, come to signify social status. This "ravening curiosity" was evident in an avarice for novelty in all its embodiments, making its appearance as the marvelous, the singular, and exotic in the literatures of romance, travel, and topography. It is expressive of a ravening culture on the rise, one in which

> luxuries are not only valuable, they are also otiose from the standpoint of utility. Early modern curiosity shared this lofty disdain for the useful. Just as political economy during this

period morally rehabilitated the luxury trade, so moralists reversed the traditional valuation of "vain" curiosity (Daston and Park, 1998, p. 309).

In an era in which curiosity is casting off its old semantic cargo as a conveyor of distraction and essential straying, by the seventeenth century a mercantile ethos grants a very decided interest in the quotidian. It is a new epistemic milieu, attending to the rare, unusual, and the particular over the traditional natural philosophical tradition, Aristotelian in method and disposition. It devotes itself to intensive enterprises and projects of inquiry across the freshly articulated disciplines, studying the ordinary in search of universal principles and characteristics.

For Daston and Park the change in fortune of the ordinary indicate incorporation of the passions in the "transition from a natural philosophy based on universals to one based on particulars required a new economy of attention and the senses" (p. 311). It is the Collector who incarnates accumulation of the extraordinary, the rare, and specimens that become valuable objects of inquiry for the new cadres of naturalists. Wonder, then curiosity, are transformed, reduced to method. The astonishment and wonder that Aristotle claimed to be the source of all desire for knowledge is subordinated to utility and material accumulation. It is a process that, for the later Enlightenment-era philosophes, becomes a form of unfulfilled thirst for knowledge in the pleasure of its pursuit, a variety of academic fox hunting.

In the taming of the passions that the early modern period devoted itself, the cresting wave of surplus curiosity was domesticated with its prodigies and monstrosities. It is an unbridled avarice for accumulation fostered by colonial trade. It is a management of the earth's resources, both material and human, rendering some of the natural world of its possessions useless. The Collector, as Pomian (1987) claims, embodies a "desire for totality," in displays of a new spirit of accumulation. Like the gamblers, credit speculators, and day traders who are its finance capital heirs, the persona of the Collector, in the seventeenth century represents a "highly refined form of consumerism, mimicking the luxury trade in its objects and its dynamic of instability" (p. 310). Pomian documents one exemplary cabinet of curiosity, that of a Dr. Pierre Borel. It enacted a racialized hierarchy in the microcosm of its chambered rooms, systematizing all things rare. Those "*curiosities*" from the Americas provided the exotic, with those from Africa the monstrous. In the catalogue to his museum of wonders, Borel had established an eccentric hierarchy of the rare, including works of God, specimens from nature, and human works. He catalogues what were purported to be the bones of a giant, "a two-headed monster and fragments of a mummy—all the ingredients of a truly fabulous and magical anthropology" (p. 46). As microcosms of the observable world, hundreds of museums of *kunstkammeren* throughout seventeenth-century Europe demonstrated an attempt to provide compendia of all that could be objectified to its panoptic Eye.[1] Museums like Borel's with its bezoars, two-headed cats, ancient weapons, and its devilfish offered the spectacle of the era's expanding largesse, as well as its abstracting taxonomy, presenting a "glimpse of nature prior to the scientific revolution. In it the play of analogies, correspondences, and resemblances facilitated the passage from the visible to the invisible" (p. 47).

The "taming of the passions" as a historical process, generated a broad cultural domestication of affect and a remapping of the uses and purposes of social space. In more modernist terms, it has constituted a continuous, though heterogeneous, social engineering and disciplining project still in progress. Altering estimations of value transformed even the deadliest of the deadly sins, avarice. Traditionally the "most foul of the all," avarice is reappraised in the marketplace. As well-being becomes dominantly associated with class interests and economic advantage, avarice is sent into the fray to harness passions that, left unbridled, would undermine a world of emergent exchange relations, prices, and abstract value. Hirschman traces how avarice was "suddenly acclaimed and even given the task of holding back those passions that had thought to be much less reprehensible" (p. 41). Beginning within sixteenth- and seventeenth-century social conditions, the passions' taming evolved as a variegated expression of the methods of an invigorated sovereign epistemology presiding in the interest of commercial and imperial projects. The commodification of knowledge entailed its corresponding, orderly objectifications of nature and human purpose.

Sweeping up continents and the futures of their peoples, colonial capture arrested perception. Perceptual activity was attenuated, suspended, paying service to accumulation. Both the metropolitan and colonial orders suppressed it, aiming attention to the seizure of material resources; their extraction disfiguring lands and their inhabitants. European rule has also meant the imposition of a regulation of phenomenal sense, of the movement and relation of bodies in space; the exact temporal order of its subjects determined by calculation. It eclipsed perception of its full spectrum of possibilities, in its tonal and tactile range, depths or rhythmic response to the world surrounding and within the colonial body. It secured the near-absolute patrol of sovereignty in the nascent biopolitical plantations and mines of empire.

As a collectable, a curiosity, the object of attention, whether fleeting or persistent, could be marveled at as a material substance, exotic artifact, or monstrosity like any of the myriad shells and mollusks in the Ambionese Curiosity Cabinet or in the possession of *virtuosi* like Dr. Borel. As objects of the mind's projected images imposed on workable matter, these curiosities were also the source for and screen for unfolding ideas, representations, and concepts. The new objects of scrutiny were often colonial subjects of empire. Bernard Mandeville (1997) was an early advocate of free trade and unfettered commercial interests through the transformation of private vices into purported public benefits. In the *Fable,* this taming of private passion and vice and its conversion into public interest is performed by the intervention of contracts and state policies:

> This was the State's Craft, that maintain'd
> the Whole of which each Part complain'd:
> This, as in Musick Harmony,
> Made Jarrings in the main agree;
> Parties directly opposite,
> Assist each other, as 'twere for Spright;
> And Tem'rance with Sobriety,
> Serve Drunkenness and Gluttony (p. 28).

Desire's propagation was the key. The regulation and the uses of the passions to which they might serve generated an intense heat that rose above moral furor. The political order comes to facilitate rather than proscribe the satisfaction of desire. Ascendant regimes of knowledge and power performed what Hirschman, like Norbert Elias (1974) before him, regards as a "civilizing mechanism"; a transformer of social relations. Its effects were dynamic and manifold, as evidenced in debates over everyday practices like sumptuary laws, the regulations for consumption and comportment. For Barbon, at issue were fundamentally the "wants of the mind." In tones through which we can hear again resonances of the appetitive figures of an unbridled curiosity, he boldly claims that

> the wants of the mind are infinite, man naturally aspires and his mind is elevated, his senses grow more refined, and more capable of delight; his desires are enlarged, and his wants increase with his wishes, which is for everything that is rare (in Berry, p. 112).

The exponential, centrifugal force of desire placed needs and their realization within an objective, natural teleological framework. Desire was now acknowledged as an indisputable element of a "natural way of things." For Berry this valorization represented an assertion that the "dynamism of human desires is self-generating" (p. 114).

In Mandeville (1997), author of the influential *Fable of the Bees*, humanity is a "taught animal" trained through the articulation and knowledge of materialist physiology. The *Fable of the Bees* was considered by its author to be an "anatomy" detailing the "invisible part of man." It offered an expertise through which the passions could be mobilized for productive industry. Passions, such as flattery, a "bewitching Engine" would act to inform the social progress of an "ideological project through which individuals were disciplined by encouraging amongst them the belief that persons demonstrated their moral superiority through acts of self-denial" (Hundert, in Mandeville, 1997, p. xxvi). In the *Fable*, "Vice nurs'd Ingenuity, Which join'd with Time and Industry" (p. 28). It was a project of domestication, prefiguring the biopolitics of the next centuries, in the calculated and "artful management" of the primal forces of the passions rationally ordered in their "strict correspondence in men and animals" (Hundert, p. xxix).

The ape of medieval inquisitiveness, the vilified *cupiditas scientiae,* exhibited as a specimen of exotica and colonial surrogate, is now mounted on the manor wall. In its useless displays, its spindrift, marvelous, and peculiar arcana, the new curiosity is conscripted in the organization, ordering, and control of colonial memory and the bright emblems of its economy. It is a distinctive moral economy structured by the industrialized dispensation of the passions in general, and the agencies of curiosity in particular, as a primal instigator. It is an economy in which the deadly sins of patristic tradition, especially the cardinal excesses of avarice and the pursuit of vain self-interest are defanged. Demonic shadows withdraw as they are now deployed to incorporate a habitus of social contracts that potentiate the power of the passions to act upon one another productively.

It is this distinctly countervailing process in which "interest" works as a pivot, hinge, or wedge between reason and the passions. It captures the momentum of increasingly restive market forces, the affective intensities of the masses in a granting of privileges that facilitate the ruthless pursuit of free trade, capital accumulation, and colonial empire. Barbon, Mandeville, Petty, Hume, and Smith propose, in their various treatises, a tendency toward a general "demoralization" of the passions. The vices, in particular self-love, luxury, and avarice are reappraised. Interest brokers a compact that would be impossible if left to the anarchic power of the unbridled passions or the weakness of reason alone. The alloy of interest, fusing the passions and an objectifying rationality, produces what Hirschman identifies as a "hybrid form" of human action. Relatively autonomous from the hegemony of Christian doctrinal practice, it provides one significant motive for the reformation of the productivity of society, or in Hirschman's words, "a new articulation of human labor and everyday life that was proposed as a solution to the inherent destructiveness of passion and the ineffectuality of reason" (p. 43). The statecraft that oversees a bridling of the passions acts through claims to enlightened cunning.

For Francis Bacon, it is a matter of "setting affection against affection and to master one by another, even as we use to hunt beast with beast and fly bird with bird" (in Hirschman, p. 22). The preoccupation with possessing and consuming knowledge occurs in this historical context with colonial subjects. The fascination exerted by the marvelous and the fixating power of wonder press curiosity into subjective forms as objects of taxonomy and classification. At the extreme, the ruthless curiosity of the colonial projects found expression in staged anxieties about cannibalism. Writers as diverse as Columbus and Montaigne document customs that are specular mirrors of colonial consumption. Cannibalism represents, in part, the limit of the appetite for absorption in extinction or incorporation of otherness. The cannibal personifies the primitive accumulation of early capitalism. It is a figure for the unprecedented trade and consumption of world markets made possible through the release of torrential productive forces, material objects and subjects in the colonial economy. The newly adventuring capitalism forms practices that profile a voracious and untamed passion disfiguring the human, difference and limits.

Francis Bacon argues for a government of states acting to "bridle" factions against each other. The worthiness of an overarching reshaping of the passions is expressed by Spinoza, d'Holbach and Hume. Spinoza (1992), who plays an often-overlooked role in the fabrication of a republican liberal polity in which human inquiry would be unconstrained by dogma, superstition, and tradition, the passions always exist in dynamic interrelation. He claimed that "an affect cannot be restrained nor removed unless by an opposed and stronger affect" (p. 136). In Hume (1965), the notion of "counterweights" is explicit in his statement that "nothing can oppose or retard the impulse of passion but a contrary impulse" that, in accordance with materialist premises, permits an "alteration of its direction" (p. 451). For Hume, curiosity is emblematic, one of the passions's "calm" expressions. It is frequently counterbalanced by its more venerable and traditional double, studiousness, a steadiness of purpose, and intent in the methodical application of a directed will. Curiosity or the "love of truth" opposes idle or useless detours from the methodical labor of thought. For Hume, hunting and philosophy are kin in their pursuit of their quarry, tarrying and diverging only in the interest of their goals. In d'Holbach the redirection of the passions is achievable by careful manipulations that do not destroy but gives them direction so as to "offset those that are harmful by those that are useful to society." For Bishop Butler, it is a return to the passion for knowledge as a counterweight: "we daily see reasonable self-love overmatched, not only by the more boisterous passions, but by curiosity, shame, love of imitation" (in Hirschman, p. 43).

Interest becomes the engine of social production. By the eighteenth century, the passions are tamed, revaluated as the "essence of life and as a potentially creative force" (Hirschman, p. 47). The calculability and constancy of the dematerialized value represented by monetary exchange, now established in minted national homogeneity, voiced a new faith in progress iconized by capital accumulation. This faith in monetary speculation and calculable futures found its rational refinement in the field of economics. Christopher Berry identifies economics as the first social science, resulting from the "natural disposition" of humans to act in their own interest. In Bethel's phrase, it is "natural for all men to seek their own profit" (p. 114). The management of the passions and reason through an aggrandizing interest found resonance in prototypes of a positivist will in a psychology which claims an innate affinity for "certainty and predictability in human affairs by virtue of Man's material nature" (p. 114). Marx (1992) paraphrases Adam Smith, regarding interest in both its economic and affective relations, as a "trucking disposition" toward accumulation. As an integral element of human nature in its unique interdependences, interest prevails over "self-love" in human activity. Marx quotes Smith, stating that

> man has almost constant occasion for the help of his brethren, and it is vain for him to expect it from their benevolence only. He will be more likely to prevail if he can interest their self-love in his favour and show them that it is for their own advantage to do for him what he requires of them. We address ourselves not to their humanity but to their self-love, and never talk to them of our own necessities but of their advantages (in Marx, p. 370).

The taming of the passions marks a signal epistemic juncture. It is a primary crossing in which money, as abstract exchange, has not yet come to dominate all spheres of commercial, social, and civic relations. Mandeville's hive is still one in which *curiosa* are held in reserve, one in which objects, in their materiality and physical presence, still prevail. Sheer material accumulation is in the process of becoming an end in itself as a signifier of luxuriance, power, and prestige. The reign of the useless object has not been abandoned. Not dissimilar from prior concentrations of power in sovereign, or in feudal seigniorial and hierarchical privilege, the ethos emerging in the seventeenth century manages not only the wealth and household economies of new mercantilists, but also propels the "civilizing" process and rationale for restructuring of social relations.

As a privileged vantage point, method is emphasized. It becomes a reflection on practices already gaining persuasive force and influence. The new epistemology of Bacon redirected older traditions of inquiry, representing the rationality of alchemy, as refined in the crafting of states, their colonial rapacity, and their handservant disciplinary knowledges. As Ann Stoler (1995) suggests, it is, if not a dialectic, then perhaps a contretemps, between a scattered imperial largesse and the metropole center. Colonial techniques, methods, and archives are brought home, collected and held in reserve alongside reserve labor and capital. Stoler quotes one of Foucault's lectures in which he surveys this process, speculating

> that colonization with its techniques and juridical and political weapons transported European models to other continents, but that this same colonization had a return effect on the mechanisms of power in the Occident, on the institutional apparatuses and techniques of power. There had been a whole series of colonial models that had been brought back to the Occident and that made it so that the Occident could traffic in something like a colonization, an internal colonialism (Foucault, in Stoler, p. 75).

Still stark in their prodigal physicality—whether tusks, cocoa, silver, cotton, or silk—it is a world where it is the having, rather than the being, of things that constitute the signs promising entry into Calvinism's registry of Providence.

A demoralization and taming of the passions inaugurates the age of empire. It makes possible a gradual transition from feudal, theocratic, and monarchial dominance to the emergence of colonial charters, entrepreneurship, and industrial and trade companies. In this process autonomous value appears. Increasingly, nothing has value in itself and the notion of independent value arises within the dictates of market relations. For Hirschman, the seventeenth century is witness to the caprice of fashion. This fashion has the ability to destabilize traditional social relations and values, volatizing them within the acceleration of market sponsorship. Free trade is encouraged to "find its own course." The shifting valences of luxury, interest, and the passions share a complex intersection in emergent capital accumulation, colonial enterprise, and pedagogy. As a revaluation of the everyday, it is the freeing of desire from the shackles of predetermined and recognizable needs for clothing, food, and shelter. Spurred by the rise of forms of pure exchange, monied transactions, values abstracted from empire, proliferate in the sanctioning of an interest in interest, surplus, and unprecedented profit for revenning enterprises like the East India Company.

The world of tamed passions is one in which curiosity is captured, in thrall to market relations. Social being begins to be knowable only by power over possessions. Colonial trophies, fixed and encased, on wall mounts, suspended, and in their *wunderkammer*, solid, robust, and often startling in their uncanniness, are readied to be taken down and dissolved into air, turned to phantoms. Credit and usury are attached to manufacture and new means are employed in distributing, maintaining, and circulating commodities necessary to the formation of a wealth losing its metallic material value. Wealth circulates incrementally, trafficking through intangible forms. It is derived and disguised simultaneously in a market economy's evisceration of slave labor and flesh for the accumulation of its ghostly surplus.

The new subject of knowledge is produced in the consequent ferment that partitions forms of practice. In the early modern episteme, a disciplinary partition was formulated along the demarcations of the courses of study of the quadrivium and trivium derived from scholastic and early university pedagogical structure. The divide between liberal, humanist pursuits and the mechanical or artisanal were premised on classical Greek models of knowledge, nature, and the arts. It presumed an unbridgeable gap between theology, logic, and mathematics and the techne of artisanal knowledges of craftwork, agriculture, and the arts. Priority was still granted to active timeless form over content. The classical, Academic model was instituted and the authoritative design for the ancient Greek society that excluded mechanical workers from citizenship and distinguished them from the ubiquity of slave labor. Techne, as the mechanical arts, the work of painters, sculptors, engineers, builders, of artisanal expertise, remained enforced for centuries, subjecting these practitioners to the privilege afforded to theology, epistemological projects, and speculative intellectual labor. Founding traditional demarcations between intellectual and manual labor began to dissolve in the Renaissance with practical alliances between theoretical *scientia* and the mechanical arts becoming sponsored by the academies then replacing universities as generators of new explorations and exploits.

The skills and procedures employed in a particular practice are comprised by any means, instruments, or tools that extend the power of the human body and that change or alter nature. Whether taking shape as a Paleolithic blade, a stylus, a hoe, windmill, clock, or microchip, *technics*, as discussed in following chapters, is marked by perennially shifting relations between human bodies and their culture's material repertoire to negotiate its interests beyond existing biological and "natural" limits. The control of technics, in its Occidental subordination to theoretical, speculative, or idle curiosities, rested on its self-authorizing claim to disinterested, timeless theoretical knowledge in the pursuit of truth. Until the late medieval era, technics was separated from *scientia*, or a *studiositas* in the scholastic community. It was cut cleanly by a curriculum that relegated *curiositas* to strayings or wantings of the mind. This separation secured the administration of a division of labor premised on a privileged and distorted reception of classical, including Aristotelian, models. With the rapid rise and organization of trades, and the emergent class distinctions stirred by colonial adventures, the taming of the passions described here assisted in a convergence of these formerly opposed modes of knowing. The usefulness of theory assured its adaptability to projects within the material world of Renaissance arts and sciences.

Emblematic of this transition were the diverse projects initiated by Francis Bacon. As Paolo Rossi (1962) claims, from the time of the Presocratics

> truth had appeared as something separate from and opposed to utility, introducing a cleavage between knowledge and operation, between logical discourse and experimental techniques. Bacon asserted instead the identity of knowledge and power, theory and practice, truth and utility, and considered opposition between these terms as exceedingly harmful (p. 151).

The mechanical model of the world of technics in the interests of the early modern savants and virtuosi displaced the production of capital; new alliances fostered by the Royal Academies and institutes served economic and political goals while advancing the new empiricist, inductive methods that placed theory in thrall to the interests of empire. An instrumental logic of invention, a hunting after the unknown worked and wreaked the extent of European nation-states' mastery over nature. Rossi cites the seventeenth-century virtuoso John Drury who stated, "as regards invention and industries, I will look principally for those which are useful to the preservation and increase of wealth through the trades and the mechanical industries, whether practiced on land or sea, for peace or war" (p. 123).

Powers of the Curious

With Spinoza, a revitalized materialism emerges, one that places the passions, or affects, within the confines of natural science. The affects are forces whose actions between bodies are reciprocal in intensity. In their relative ability to influence others, the affects are corporeal powers, increasing or decreasing the body's degrees of motion or rest. Curiosity is always a situated, reciprocal relationship between bodies. An exchange of energies is always present between human agents and the material processes and structures that comprise the bases for the natural and organic worlds. Regarding it as a passion of collective mind affords a unified view of the formative principles of human nature and agency. It is premised on the powers of bodies—human, animal, and inorganic—in the continua of their active interrelations. It surpasses the persistent fault lines of all those who accept an ontological or epistemic split, the infamous "mind–body" problem. In this perspective, affects and such passions as curiosity are never separated from the physical forces and processes of nature. It is not divided, as ontological substance, between states of mind, thought, and corporeal experience. For Spinoza (1992),

> if anything increases, diminishes, helps or limits our body's power of action, the idea of that thing increases, diminishes, helps or limits our mind's power of thought (III, p. 137).

Curiosity as a passion and power to act, in a Spinozist conception, entails dynamic relation of bodies to each other, of worlds within worlds of mutual influence and composition. His is a monistic, systemic principle in which the kinetic powers of bodies act in ongoing reciprocal activities within quantitative degrees of force and qualitative transitions of state. Bodies, in their effect on one another, act in perpetual tension and regeneration. Their relational dynamics are transformative and mutable, always implicating simultaneous covariation, which also oscillates in thought, concept, and representation.

Spinoza provides us with a significant key to a perennial epistemological war between the affects and reason. His well-recognized rationalist tenets provided a basis for his defense of the Republican partisans of free thought and intellectual curiosity in the face of their looming defeat to Calvinist reactionary monarchial interests. Spinoza resisted a remobilization of theological and political power. The reactionary forces opposing the then fragile Dutch parliamentary state under the de Witt brothers (who were murdered in 1672) justified much of their rationale to reclaim power for a despotic monarchism through metaphysical doctrines. Spinoza belonged to a small coterie of freethinkers and intellectuals who supported what may have been Europe's first tentative experiment in forming a liberal democratic state. His materialist premises, in both the *Ethics* and the *Political-Theological Treatise*, testify to an unparalleled reformulation of the grounding principles of ontology, ethical interest, and the nature of social organization. His heterodox premises were heretical enough to earn him a double banishment, first by official excommunication from the Amsterdam Jewish community, and later through that city's increasing intolerance toward the advocacy of secular, atheistic free thought.

In terms of social contract, Spinoza's monistic conception of the unity, without identity, of mental and intensive with physical and extensive substances, led him to draw their correspondence in the realm of political rule. This parallel or homology between substances and forms of social organization, and its corollary correspondences between body and mind, allowed him to think through the arbitrary nature of the justification for natural, or absolute right to domination. He argued instead for a dynamic interchange between rulership and the ruled emphasizing the pivotal, if ambiguous, role of the passions and power of the "multitude."

Processes for the taming of the passions are contingent on internal developments within their respective and distinct social collectives. They bear witness to relative degrees of compensatory release through forms of social agency and practice. Any taming of the passions takes root within a qualitative historical, experiential continuum that adjusts and accommodates their purposive release. The suppression and maintenance of specific economies of curiosity—whether during the rise of natural science from its "natural philosophical" antecedents, as in the trial and recantation of Galileo, or in the still-infamous hanging of Bruno in 1600—exists in reciprocal determination with more liberal economies of affect and agency. Suppressive discourses, including states of exception to constitutional or parliamentary democratic polity, through various mediating forms, such as those deployed against citizens and immigrants in the permanent "war on terror," cohabit with social imaginaries that are never absolutely repressed. Surveilling social powers, whether theocratic, monarchical, or the instrumental rationality of market and social democratic states, ignite fires of all sizes. Rebellions and insurrections, as illustrated by the Events of May 1968, discussed in Chapter 4, manifest the collective desire for revitalized forms of life.

Theoretical curiosity is social passion. It circulates with the power of bodies to act upon each other in the interest of realizing social potentials. Historically, it has most often vied with and made compromises with the stability and continuity of existing orthodoxies, at times engaging in combat with presiding clerisies and dominant discourses. The passion for knowledge has challenged closed economies of inquiry and innovation—its taming channeling it into conservative reproductions of existing symbolic and material practice. This cynically encourages false degrees of freedom for curious minds and energies within the encompassing forms of social and individual interest. All artists during the Renaissance required patrons who might or might not support for innovative ideas and techniques. Parliamentary reforms led to a change in sovereign power but did not lead to what we would consider now the functioning of participatory democracies. In Spinoza's terms, it is the creation of new unities and forms of collectivity through the formation of "common notions," as voluntary contracts, compacts, or accords. These constitutional arrangements are able to subsume individual passions, including curiosity, to the benefit of greater social interests that promise a fuller realization of human powers.

The transformation of individual passions into collective imaginings represents for Spinoza the melding of individual "natural right" into civil right. This conversion is necessary in some manner for all communal structures of sociality to cohere. The interests abetted by the compacts of civil rights are meaningful as far as they make mutual aid in daily life function with a minimum of conflict and physical violence. When they cease to serve the greater interest, or when competing elements of the polity find they are not benefiting from participation in the social contract, as citizens of a republican state, that contract becomes unstable and can be sundered. What is significant here is that the passions, never entirely quelled and submitted to an overarching social structure or function, remain the driving force of collective life.

Most importantly for this study, Spinoza's propositions allow for a reevaluation of curiosity, an active, elemental principle within the spectrum of the passions, agency, and affects. It is a natural, physical process, and a prowess common to humans and other sensate creatures. As a passion, curiosity partakes of the fullness of Spinoza's naturalization of mind; it is the locus of productively rational, intellectual, and affective powers to affect the actions of other bodies and minds. As Gatens and Lloyd (1999) explain, "human knowledge is embodied knowledge and different ways of knowing imply correlative ways of being" (p. 104). As a catalytic element of lived embodiment, of both social and individual existence it is not finally reducible to intellectual or psychological processes. Anticipating aspects of philosophical inquiry in the twentieth century, Bergson and Deleuze view it as an affective-cognitive intensity or singularity within all cultural practices, including the technosciences. Consciousness, for Spinoza, in contrast to Descartes, is an effect, an afterimage of the historical play and agon of the passions.

The Encyclopedie and the Philosophes

Intellectual curiosity in modern thought took on a particular contour within the momentum developing within the culture of the natural sciences of the Enlightenment era. Refashioned for the Republican subjects of a secular and materialist era, it is one of the unsung avatars for an episteme marked by the Encyclopedists. It is an age in which wonder, astonishment, and curiosity will give way to the careful, empirical search for universals and critique. Kant's formulation of philosophical critique as a genre delineated a cognitive map of the limits of knowledge, judgment, and understanding. Intellectual pursuits and the "faculties of the mind" participate in the formulation of the distinctively new forms of subjectivity made possible by the dramatic transformation of the religious, moral, and political tableaux. Carlo Ginzburg (1976) finds it significant that Horace's motto "*sapere aude*"—dare to be wise—was a shibboleth for the Enlightenment's surpassing traditional boundaries of knowledge. In answering his own question, "What is Enlightenment?" Kant answered "*sapere aude*"—dare to know (p. 41).

The means, strategies, and degrees to which a culture directs, deflects, or leaves inquiry relatively unbridled remain key questions about the limits to the power of questioning subjects. Gradually reprieved in early modernism from its vilifications as the "eighth deadly sin," *curiositas*, was no longer a form of purposelessness. With the interests of a European secular and humanist culture, aspects of its classical theoretical value were retrieved. By the time of Bacon, it was to become an experimentally sanctioned signifier, replacing, while seizing upon its classical status as "reposeful and bliss-conferring contemplation" (in Blumenberg, 1985, p. 385). In its Enlightenment formulations, as inscribed by the Encyclopedists, it was both challenged and valorized as an "unrestricted cognitive drive." What Blumenberg identifies as a reinvigorated "theoretical attitude" comes gradually to restore theoretical and intellectual curiosity's stature.

In the Preliminary Discourse to Diderot's *Encyclopedie*, D'Alembert (1975) declares "curiosity is a need for anyone who knows how to think." In the great totalizing tome of the Enlightenment, the pedant clarifies for us that "the mere fact that we have occasionally found concrete advantages in certain fragments of knowledge, when they were hitherto unsuspected, authorizes us to regard all investigations begun out of *pure curiosity* as being potentially useful to us" (p. 17). In the fashionable salons the quite sociable persona of the new *philosophes*, curiosity is a pleasure inherent to the pursuit of the goal of a true and manifest knowledge. It is useful knowledge, crafted and supplied to fit the purposes of statesmen, naturalists of the empirical sciences, explorers, and imperial exchequers. It is no longer the purely enjoyable knowledges nor the *curiosities* of the collectors and their cabinets of wonders, lacking all luster of useful truths. D'Alembert speaks for the philosophes and their eighteenth-century audience, pronouncing curiosity a reward, with utility serving as its pretext.

The *Encyclopedie* is a textual font for a will to systematic knowledge. Authorized in the name of natural philosophy it boldly attempted to set Enlightenment's intellectual compass. It sets its points of reference in opposition to the prior generation and influence of Descartes, Newton, Leibniz, and Spinoza. Its purpose was to supplant the speculative rationality of these earlier theoreticians, lovers of a "spirit of system" in itself with systematic empirical pursuits. Intellectual curiosity, however, was ambivalently represented by the philosophes. It was condemned again by members of this totalizing movement toward universal knowledge. Voltaire participated in a generalized moral ridicule of practitioners of curiosity who deployed a theologically and metaphysically unfettered discourse. As Blumenberg argues, for the philosophes, "curiosity alienates man from his true basic problem, which is a problem of behavior, not of knowledge" (p. 405). In Voltaire (1961), this is explicit in his cry: "Oh man! this God gave you understanding so that you should behave well, and not to enable you to penetrate into the essence of the things He created" (p. 14). Voltaire retained a moral sanctimony not all that distant from medieval interdictions of curiosity as one of the forms of purposelessness, of *acedia*, a lapse from the devotional life's "essential centering."[2] A new restriction of its limits and value for research was imposed in condemnations now arising not from patristic and moral discourse, but through the invocation of a fulgent empiricism of mastery over nature.

For the philosophes curiosity represented thought without system, random perceptual sensation and speculation lacking method and empirical rigor. In Voltaire's

Philosophical Dictionary, it is a "'plucking feathers from sparrows." Like Plutarch's assault on it for its alleged voyeurism, it is an eavesdropper's attention: a glance to the side, noticing the clothing of a stranger, the mail a neighbor receives, the physical attention and affection exchanged between acquaintances. Voltaire, an active enthusiast of Newton's mechanistic and unifying principle of phenomena, dismisses it as a

> passion alone which produces the immense attendance at public executions for when one of the academicians of Paris tried to get within the enclosure to examine what was passing more closely, and was forced back by one of the guards, "Let the gentleman go in," said he, "he is an *amateur*, that is to say, he is inquisitive." It is not from any malice that he comes here; it is not from any reflex consideration of self, to revel in the pleasure of not being himself quartered; it is only from curiosity, as men go to see experiments in natural philosophy. Curiosity is natural to man, to monkeys, and to little dogs. As to men, you know how they are constituted. Rome, London, Paris, all pass their time in inquiry: what's the news? (p. 74).

In the *Encyclopedie* itself, some entries disdain these masters of worldly knowledge, cast in the direction of the spirit of unconstrained inquiry. In the section entitled *Soul*, for instance, we are forewarned "how unfortunate, for an insatiable curiosity, for an unquenchable thirst after well being, that we are thus ignorant of ourselves" (Diderot & d'Alembert, 1751–80, p. 324). Their concerted systematic approach notwithstanding, perception and theoretical pursuits together were reduced to instrumental purposes, made equivalent as raw materials of empire.

This will to systematize all available knowledge is founded on the never-innocent or disinterested empiricism of method and orientation. As John Willinsky (1998) illustrates, "French efforts to gather up the world's scattered knowledge, to expand the range of its sensations, named the intellectual debt owed to imperialism" (p. 73). Among these attempts were the methodical instructions, techniques, and practices necessary for the proper industrial appropriation of colonial material largesse, including the manufacture, curing, and treatment of manioc, indigo, tobacco, cotton, and sugar. In this period, curiosity becomes a raw material itself in the self-authorizing and sanctified aggrandizement of the bounty of an apparently limitless plunder of the earth.

As the target of a signature sardonic wit, an untamed appetite for knowledge becomes the theme of Voltaire's extra-worldly *Micromegas* and the *Diatribe of Dr. Akakia.* In the philosophical tale *Micromegas*, (2001) beings visiting Earth in search of utopia have seventy-two senses. One of the novel's characters exhorts the multiple-sensed, yet bored and craving visitors: "you are nothing but a curiosity seeker—you have never experienced love." For Blumenberg this expresses a more generalized antithesis between curiosity and happiness in Voltaire's *Contes Philosophiques* and among the philosophes. For Voltaire, "even a 1000 eyes could not satisfy curiosity and eyes that were a 1000 times better would not improve the prospects for satisfying the unrest of the cognitive will" (in Bluemberg, p. 406).

Voltaire directed his intense parodic powers on one paragon of an emergence of untamed inquiry, Pierre Louis Moreau de Maupertuis. The *Diatribe of Dr. Akakia* uses the life and researches of this natural philosopher, astronomer, biologist, experimenter, and adventurer as the basis for its scornful derision. Blumenberg devotes deserved attention to Maupertuis, claiming that he "stylizes the prototypical role of the functionary and hero of *curiosity*" (p. 408). Maupertuis is an exemplar of an unconstrained epistemic will to knowledge and power pertinent to contemporary ethical issues emerging from technoscientific inquiry. Maupertuis was among the first to advance research in the natural sciences that would explore "artificial unions," actively encouraging, in his "*Lettres sur le progress des sciences*," experiments on the living, including institutionalized charges of the state such as prisoners. Prefiguring both Shelley's Frankenstein and the specter of eugenics, his experiments ushered in "the breeding of the kind of curiosity that previously were only collected—intentional, methodical, artful breeding, by means of artificial unions" (p. 411).

An avatar of what Nietzsche declaimed as a "ruthless" curiosity, Maupertuis's polymathic investigations were also forays across the limits and surfaces of existing knowledge into the hybrid, heterogeneous, and monstrous. He proposed the artificial union of a "bull with a she-ass," and other "forced procreations." Maupertuis's research adventures were wildly inclusive, including proposals to disclose the unknown materials inside the earth, attempts to detect unseen stars by use of new telescopes, and foraging the terrain and fauna of Patagonia. He challenged all existing systems and disciplines of knowledge for their superficiality and resistance to disclosing the unknown. These adventures claimed to test the limits of human knowledge, art, and power over nature. As with Francis Bacon, an inspiration for the period, his Promethean will to unfettered knowledge advocated the use of violent means if they would advance the pursuit of theory, and plumb the depths of the knowable. This is illustrated by his proposal to use gunpowder to blast through to the secrets of the pyramids. This is a singular curiosity now redirected from an "interest in discoverable curiosa (curiosities) generates them itself. This is what is new" (p. 412). Maupertuis regarded his own projects as forays into theoretical curiosity. Theoretical curiosity is pursued in an empirical abandon, disembodied from the effects of its power in and on material existence. In his *Letter on Cosmology*, this taming of the limits of existing knowledge is explicit:

> when I consider the narrow limits within which our knowledge is confined, the extreme desire that we have for knowledge, and our incapacity to instruct ourselves, I could be tempted to believe that this disproportion, which exists today between our knowledge and our curiosity, could be the result of a corresponding disordering event (p. 416).

Curiosity, in this optic, is a process of dislocation, an untimely interruption a radically distanced mind. Maupertuis's advocacy is Faustian, a boundless rending of the existing limits of disciplinary knowledge. His persona is representative of the perpetual tension between the ideals and terrors associated with the formidable powers of this singular passion.

The *Encyclopedie's* article, *curiosite,* accentuates the passions of the intellect. In depersonalizing its inherent dynamism from its actual practitioners, the author, the Chevalier de Jaucourt, asserts its pervasive presence in objective instances and encounters. He attributes it to a certain instability, an *inquiete,* reflecting in part its ambivalently rehabilitated status within the aspirations and ambitious projects of both empirical science and the Encyclopedie philosophes themselves. It is described as a desire independent of any of its possible objects or personifications. Contrasted with the article on the *curieux,* the collectors, savants and amateurs, Jaucourt's significantly larger contribution affirms curiosity's elevated, if again ambiguous, position within the Enlightenment pantheon as an inherent, restless desire for knowledge. While *curieux* maintains its semantic referent in identifying a distinct social caste and category, *curiosite* has become a universal aspect of creative intellectual endeavors and sustained methodical inquiry. Its ethical charge usually neutralized, it becomes autonomous from the objects of its encounters. Jaucourt's essay provides a new reading of its pertinence to everyday pursuits, one in which "good objects of this desire for knowledge are defined in distinctively philosophe terms, as one that will lead to progress and the improvement of society" (Kenny, 1998, p. 76). Material objects and their cataloguing are distinguished from intellectual knowledge, possession from pursuit.

Jaucourt's curious desire has no preestablished origins. For Kenny it has become a "free-floating desire which circulates and flares up here and there, rather than a fixed characteristic of certain people" (p. 78). Anticipating tenets of contemporary theory—in particular Deleuze's rhizomatic thought, discussed in Chapter 7—this formulation resists the disembodied epistemology founded on Cartesian and rationalist grounds of inquiry in favor of a more materialist account of mind. For the philosophe, *curiosite* is a desire that arises from a "friction between the soul and sense perception of objects." As a desire that does not even "originate in people themselves, it is not a movement out from subject to object (as it had traditionally been conceived) but instead it arises as a fold or interface between subject and object" (Kenny, p. 78). This is a premise suggestive of Stoic thought as well as aspects of phenomenology I discuss in the context of contemporary philosophical critique.

A double displacement is at work in the valorization made official in the august pages of the *Encyclopedie.* First, its ethical valence is erased, as is its association with the material collection of objects, especially evidenced in the plurality of interests of collectors in contrast to the self-consciously methodical curiosity of the philosophers themselves. As practitioners of the positive aspect of this desire, collectors are exemplars, paying heed to one intellectual pursuit or object at a time with diligence and care. Secondly, there is a displacement of knowledge from any pregiven relation to objects in general, and especially away from ascribing to it any universal attribution. Curious desire is not containable within the provinces of subjectivity to which it is reduced and debased by idle accumulation. It becomes the subject of an irreducible excess that rebounds within uncertain "interfaces" in dynamic and contingent relations.

The *philosophe's* rigorous shedding of the long moral shadow and censure from *curiosite* has come to define, in large measure, its contemporary reception. Its moral value now dependent on its objects, was, and remains, a reflecting screen for technicians and scientific inquirers themselves. Kenny claims that with the *Encyclopedie,* curiosity became the "universal starting point for the sciences" (p. 78). Its moral rehabilitation justifies its rejuvenated epistemic role in claims for truth. Its decline begins, however, at this very apogee. Conceptualized as a "brief, initial phase in the acquisition of knowledge," (p. 78–9) it remains associated with a certain superficial encounter, a transitional and temporal feature of the mind's workings. It has become an individuated, psychic desire that cannot last, that must yield to inductive method in rationalist pursuits of universal processes and laws of the natural world. The mathematical models of classical speculative thought were fashioned for the exponential development of mechanical models and processes intensifying modes of production and capital accumulation. Despite its tentative celebration, Jaucourt does acknowledge its enigmatic qualities. He indicates a paradox in this seed of thought, one that transgresses the dual question of limits and limitlessness. The Encyclopedie philosophe comments on a perduringly enigmatic *curiosite*, which expresses a "strange habit of naturally exceeding its own natural boundaries and those of our intellectual capacities. There is something about it that stops ceding, humbly, as it should, to attention" (p. 80).

Ruthless and epistemic, "idle" or commonplace, variants of the passion of curiosity intertwine, participating in the fabrications of daily life. In the brief historical profile sketched here curiosity gained a respected place at the table of the arbiters of reason. It played out its new status within the empirical sciences alongside the card games and conversations of the salons. Refigured, it led a new life as a double agent, bearing its ruthless and epistemic, Orphic and Promethean conceptual personae. The empiricism inaugurated by Bacon's New Science redeployed the place of the mind in conjunction with the work of the hand. Intellectual and manual labor were allied, scientia and techne merged in joint ventures of utility and mastery over the world's objects and subjects. Empiricism, never uncontested nor untheoretical, permitted an oscillation in the value of research into the tangible and material potentials of the natural sciences and their emerging applications to the biopolitical management of the passions.

Leading its multiple lives within changing epistemic cultural patterns, this singular passion animated its doubles and dualities. For Foucault curiosity constituted the strange appearance of "man" as the prized modernist object of knowledge, incarnated as an "empirical-transcendental doublet" (1994, p. 318). It animated the visible forms that could cohere as discursive disciplines and objects of knowledge. The taming of curiosity, amid the gradual cultivation of civility, as documented by Norbert Elias (1974), masked the manipulations of Enlightenment knowledge and power and its burgeoning empires, both domestic and exterior. In the early modern period it demarcates both a sign of human endeavors in their potential and actual transgressions of moral probity and discipline as well as serving as an objectification of the passions and interests. Kenny claims that for the early modernist sensibility of the curiosa and the philosophes alike curiosity's "boundaries are never impervious, its closure never complete. Disputes over boundaries of curiosity were particularly controversial and highly charged" (p. 30).

Never isolated, epistemic and ruthless, idle and intensive forms of inquiry intersect with speculation and interest. As documented at the level of its etymology, curiosity has always been an unstable concept. As an "open ended, incoherent cluster of countless meanings and narratives that have been ascribed to it in different periods" and contexts, curiosity has been exemplary as a supple medium and mediating human form of agency put to the ready in "games of power and legislations" (Kenny, p. 31). The robustly empirical employment of the passions in the interests of states and empires in their management of trade and accumulation, as traced here, offered a material complement and support to the transcendent, theological guarantees to their dominion and sovereignty.

The passion for knowledge, inquiry, and the pursuit of what I consider—in the discussion of the place of the everyday in Chapter 4—a critical curiosity will always worry the vulnerable edges and closures of any social totality. It is the power to think and act in the interest of collective life lived otherwise, an open-ended questioning that imagines a plurality of futures.

Notes

1. An interesting historical precedent can be illustrated in the origins of the Ashmolean collection. Established in Oxford in 1693, it was formed on the basis of the material provided by one of the first English cabinets of curiosity. John Tradescant the elder, a garden designer employed by King Charles I, collected the "Biggest that Can be Gotten," rarities obtained from returned travelers and adventurers. The collection, inherited by his son, was left to Elias Ashmolde (1617–92) the author of the first systematic catalogue collection. As James Hall reports, "Ashmole was a true 'renaissance man': an expert administrator and lawyer; collector of coins, books and manuscripts; historian of the Order of the Garter; founder member of the Royal Society." In Hall, James. (2009). The Politics of Display. *The Guardian Weekly*, 18,12, p. 48.
2. Commenting on an earlier period of European culture, Stephen Greenblatt observes that even as early as the fourteenth century curiosity bore the imprint of more than one master and intent. Greenblatt discerns a "heterodox" theoretical curiosity at work in the context of Mandeveille's adventures, which reflect a "refusal to grant the universal authority and ontological priority of Church orthodoxy, a set of interests that spin away from the centrifugal force of Christian beliefs." Greenblatt, S. (1991). *Marvelous Possessions: The Wonder of the New World.* Chicago: University of Chicago Press, p. 46.

Chapter 3
Pedagogies of Curiosity

A critical questioning and questioner stands like Thales of legend. In awe not only of the vastness of the night sky but also the manner of our bearing, a restless questioning is a caring for the inquisitive interests of contemporary cultural and scientific practices. In this and the next chapter I wish to profile ways in which distinct practices, in particular pedagogies, brought into critical question, let us place accent on the nature and experience of everyday life, its rhythms, order, and potential for transformation. This inquiry into processes of inquiry itself begins with acknowledgment of it as an abiding desire for knowledge for the shaping of the present and future worlds.

An uncritical empiricism would have us not think the qualities of the primacy of the sense world into which we are always engaged. Dualistic, objectifying representational thought, ubiquitous and long challenged, in both the sciences and the cultivation of the modern subject, eclipses the perceptual coils shaping our embodied experiences. As Renaud Barabas (2006) comments, any critical exploration "must pass through an inquiry of perception" (p. 11). I attempt here to historicize and place the inquiring mind back into contemporary practices and pedagogies in making a critique of a usually unreflective identification with the otherwise individualized and isolated aspects of a purportedly universal "human nature." My emphasis here is to abandon any a priori reductive psychological or philosophical premises regarding the critical and imaginative faculties of the mind. It is my interest to put into serious play the chameleon of a critical curiosity; one once made present for what a robust pedagogy of curiosity and its agents could recreate in the production of knowledge in a continual (re)-education of the senses.

My mapping of the unstable territories of curiosity is not intended to portray human intervention as ever merely motivated by the convergence of it as some purely vitalist energy or desire. This project's explicit purpose is to demystify the ambiguous status theoretical curiosity holds in contemporary life. I choose a historically nuanced identification, advancing a socially enabled critical curiosity as a generative practice and questioning of the everyday. This openly interested approach is partial to strivings for recognition, autonomy, and the equitable redistribution of the earth's natural resources. This distinctly "interested" accent to theoretical work of all kinds is intended to serve in this project's consideration of the unchartable

M. Zuss, *The Practice of Theoretical Curiosity*, Explorations of Educational Purpose 20, DOI 10.1007/978-94-007-2117-3_3, © Springer Science+Business Media B.V. 2012

continua of experiences and the primacy of the sensible and perceptual in all their variety and virtuality of human paths of inquiry. I will give it one location, starting with a sketch of the place of inquiry within the context of the experiment and experience of the new republic of America and its contested role within pragmatism, one of its own most germinal inventions.

On American Utility

Presiding ethical values place the ambivalence of writers like Freud toward the expression of curiosity into the North American context. Posnock (1991) argues that

> even in late 19th century the status of curiosity remained polemical, especially in utilitarian cultures like those of England and America. In both countries curiosity remained an attitude and a mode of being saturated in both medieval and Christian interdictions and the Baconian revaluation of curiositas (p. 38).

In the competitive productive climate of nineteenth-century industrial capitalism, utility and purpose gain priority over theory and speculative inquiry. Captains of nineteenth-century industry and finance, and founders of the first American research universities, like Johns Hopkins, were not praised for their curious intellects and spirit of inquiry. Opening frontiers to expansion and exploitation, Morgan, Rockefeller, Astor, and their heirs today in digital, genomic, nanotechnology industries deploy individuals and legions of explorers, adventurers, and inventors in their research and development efforts and their need for ever-expanding markets. The Puritan legacy and lineage still held sway in its ardent practical attitude toward the purposefulness and consistency of human moral action in the world. It also generated the long-held suspicion of intellectual, high-brow culture and authority. It was John Cotton, for instance, who declared "the more learned and witty you bee, the more fit to act for Satan will you be" (in Hofstadter, 1966, p. 46).

This legacy of moral responsibility resonates strongly in the Anglo-American context where an abiding anti-intellectualism has long found its hardscrabble soil. Intellectual inquiry is implicated in the enduring ambivalence, rocky reception, and more than occasional acts of censorship that critical research activities have often received in America. It shares the travails of those relegated and segregated, by intent or social effect, to a purportedly elite intellectual class. In Richard Hofstadter's (1966) classic account these "elites" have often come to be known as intruders. Their efforts at reforming or reconceiving the traditional moral and ethical compass points of the social world have often met with suspicion, Biblical reckoning, and resistance. The rise of the Tea Party in American culture illustrates again a deeply engrained hostility to "elites." Epistemic inquiry is the more than occasional traveling companion of intellectual activity in American life in its continual struggles to reinvent itself. Open intellectual inquiry bears the brunt of invective, ranging from irrelevance to apostasy and subversion. Conservative and right-wing vitriol has been spent on equating free intellectual inquiry with a demonized liberalism and cultural pluralism. Rebellious and recalcitrant, intellectual interventions can act as stubborn goads to the American frontier psyche in its protean reinventions of identity. Traveling and trafficking, but not incarnated neatly by any specific community of intellectual agents, it repeatedly confronts traditions and disciplines of thought and moral rectitude.

As Hofstadter reminds us, intellectual critique and engagements have their own history of pieties and dogmas; it is also a historical narrative of a flight into the play of the mind and the pure pleasures of its own critical pursuits. When acting in relative autonomy from the moral and ethical sanctions or the demand for the usefulness of all knowledge, these can amount to indulgent, ludic forays of deliberately theoreticist and speculative intellectuals, scientists, and mavericks in their often-privileged ambivalent and ironic posturing. A demand that all knowledge be useful, at least as potential reserves of conceptual capital, has never worn off its nickel-plated patina. Invectives against intellectual work find expression in ongoing diatribes about both the essential utility of all possible knowledge and its purposefulness in a greater plan of Providence, with ultimate claims to divine and "intelligent" design and prohibition since Eve's gazing eye in Eden. Curiosity is blindsided, a hired hand, recruited in war, the management of the new body politic, and used in the conquest of markets. "Think tanks" like the American Heritage Foundation, American Enterprise, Cato ("Individual Liberty, Free Enterprise, and Peace," which is funded by the Koch brothers and their FreedomWorks tea-party network), and the Manhattan Institutes, for instance, serve to produce cultural capital, formulating strategies, rationales, and the planning for a ceaseless making and maintenance of new empires of privatization.

The conflict over the degree to which knowledge must provide utilitarian goals can be traced to colonial American priorities. With the founding of American Philosophical Society in 1743, which reproduced the premises of the Royal Society of London, utility was made its explicit objective. In Benjamin Franklin's statement of purpose the Society would work toward the expression and promotion of useful knowledge from the

> hints, observations, received from men of speculation pursued and improved...[to] produce Discoveries to the Advantage of some or all of the British Plantations, or to the Benefit of Mankind in general (Machlup, 1962, p. 10).

The core membership of the new society consisted of a physician, a botanist, a mathematician, a chemist, a mechanician, a geographer, and a general "Natural Philosopher." Imitating London's Royal Society, the colonial institution would follow its course in studiously refraining from "meddling with Divinity, Metaphysics, Moralls, Politicks, Grammar, Rhetorick or Logick" (p. 11).

Within the American pragmatist tradition the notion of an open-ended realm of the knowable and possible recurs. Posnock, writing in the cultural context of William and Henry James, argues that the pragmatist tradition generally "welcomes an unfinished, growing universe." For William James (1978), we live in a "tramp and vagrant world" that "leans on nothing" (p. 124). Posnock documents the recurrence of premises that locate curiosity as the place of anomaly and irregularity in relation to structures and systems of knowledge and experience. Worrying Peirce's "habits of embodiment," curiosity is a point of departure for embarking on sustained study and research. It is often reflected upon as an experience of a "'release' that does not fit; neither a symptom of repression nor a substitute for action, its libidinal energy escapes the tyranny of metaphysical dualisms" (in Posnock, p. 50). This premise will also recur throughout journeys of intellectual and technoscientific projects. In contemporary physics research, this is expressed by Richard Feynman (1999), for whom critical thought is experienced as a sustaining of the "pleasure of finding things out". It is the "kick" in the process, propelling discovery: "the thing that doesn't fit is the thing that's most interesting, the part that doesn't go according to what you expected" (p. 14).

For William James (1979) an Aristotelian wonder that sets the impulse to know and philosophize into motion must be discarded "if thought is not to stand forever pointing at the universe in wonder" (p. 65). Despite his continual emphasis on process, risk, and change in an "unfinished universe," James sought in pragmatist method a way of resolving otherwise-unending disputes in philosophy. For James, "man always wants his curiosity gratified for a particular purpose" (p. 165). Demonizing a "diseased" dispersal of intellectual energies, James is succored by a thriftiness, prudence, and utilitarian orientation that Veblen found particularly prevalent in England and America. Pragmatism, in engaging experience, collective and individual, is a method of skillful and shrewd management of the mind. Natural scientific inquiry jettisons the distractions of everyday life in which the "eyes are fixed on vacancy" and in which we submit to an "empty passing of time" (James, 1983, p. 382). In James's articulation, the mind is only "efficient" insofar as it is capable of "narrowing its point of view. Otherwise what little strength it has is dispersed, and it loses itself altogether." It is a position at odds with James's espousal (1979) of a wild empiricism, his "pluralistic" universe of discovery (p. 65). This dispersal, akin to medieval *acedia*, is again apparent for James in the "vice of subjectivism," as well as in literary romanticism. It represents a threat to the freedom of the will, and like Freud's criticism of Leonardo's overabundant curiosity, discussed below, it vitiates intellectual, libidinal, and creative energies that could be put into practice. An idle or "interminable" curiosity is a "fatalism" that keeps us caught within a "labyrinth" (1979, pp. 132–33).

In contrast to William James's condemnation of the idle, useless, and "interminable" curiosity he found exemplified in his brother Henry, Thorsten Veblen (1918) praises it. As Posnock recounts, Veblen thought James's disparagement of the spirit of an independent inquiry to be pragmatism's "blind spot." In particular, he criticized the basic pragmatist tenet that asserts intelligence as fully functional, available as a "reasoned line of conduct looking to an outcome that shall be expedient for the agent" (p. 6). Information and its continual pursuit can be ends in themselves. Veblen is singular among his cohorts in arguing unabashedly for the place of the idle, for ways of acting and thinking, like the purported "Pueblo mythmaker" whose consequences are unintended and irrelevant. He asserts that idle curiosity, closely related to the power of play in human life, is the instinct that has motivated mankind "toward a more and more comprehensive system of knowledge" (p. 9). Appearing in myth making and legends, idle, speculative curiosity is also the spur to modern science. Veblen adapted the premise of his mentor, C. S. Peirce, the originator of American pragmatism, in this regard. It was Peirce, usually known for his "semeiotic" logic, who proclaimed that "true science is distinctively the study of useless things—it is destroyed as soon as it is made an adjunct to conduct" (Collected Papers, 1931–1958, pp. 1, 76). Veblen's argument was aimed against a wholly functional and objectifying stimulus determined psychological reduction of social knowledge and practice. He intended to escape the dichotomy becoming enshrined in the utilitarianism to which pragmatism was being kept hostage at the dawn of the Progressive Era. This was the core dualism that James opposed as he tirelessly advocated a scientific and epistemological curiosity that would aid practical questions and activity. Veblen vigorously defended what for James's was the uselessness of an idle and "interminable curiosity" concerning itself with the supposedly ephemeral nature of the sensory and the perceptual.

As I reviewed in the classical and medieval contexts, "idle" curiosity was a signifier for the realm of perception, of the senses in general, and of the visual in particular. It has traditionally been opposed *to theoria* or *episteme*, forms of ideal, speculative and non-sensuous knowledge. In its provenance from the Greek Academics through to various modern reductions, it remains usually occulted, derivative at most, of its epistemic master. As knowledge's self-pursuit, in the tracing of the anomalous, unusual, the peculiar, idle curiosity it is frequently construed as no more than a sometimes not so innocent dalliance with the perceptual givenness of things, especially with the novelty of the merely contingent or accidental. C. S. Peirce was robustly contrary, speaking against the grain when he remarked that science's purity resided in idle inquiry, in its essential uselessness.

This "idleness" of the mind gains stature when acknowledged as a constituent "cognitive drive." Originating now from an epistemological harbor, a launch for the mind's expeditions and exploits, it is recognized as a primary point of origin for the development of systems of thought, concepts, and representation. In discarding its passional interests, it attains the prestige of sober philosophical, scientific inquiries, such as those initiated variously by Leonardo, Maupertuis, Darwin, and von Humboldt. It comes to act as a transit point for the arrivals and departures, the presentations and vanishings of cultural objects of study. It marks the production of surplus value to forms of inquiry enmeshed in trade, surveilling potentials, in economically and culturally significant forms of development.

Rarely unfettered, this idleness can appear anywhere. It discloses interrelations between disciplines and discourses. It assists in the building of the edifice of research paradigms as well as in their demolition. It is expressed in the necessity to sense, detect, and pose problems that bring new questions and approaches to accepted doctrine and method. What is often witnessed in the development of a modernity's research programs is a reduction of complexity and ambiguity. This occurs in a willful turning from open-ended inquiries to intensively instrumental methods and the amassing of sets of data, a shifting of gears to what, borrowing from Nietzsche, could be characterized as a more "ruthless" aspect of any research that purports to be disinterestedly epistemic. This is the Promethean objectifying violence that, with its instrumental reasoning and technological apparatus, seizes the empirical, forcing the creation or, for Gaston Bachelard (1938), the "invention" of its own epistemic objects. An empirical, disparate plenitude is made to reflect the rationalism or realism of its research community. The "natural" world is pictured and captured in representational, often positivist, pattern, ready symmetry, and replicable order.

Capital investments in academic research have risen exponentially since the 1980s. Corporate alliances with university-based research have grown 8% each year since 1980, amounting to a total of nearly 2 billion dollars in investments each year. This financialization is directed to the construction of industrial parks, venture capital funds, early-stage drug development operations, and the securing, patenting, and licensing arrangements through technology transfer offices. The decisive turn toward an entrepreneurial symbiosis between corporate and academic resources in research and development was established with the 1980 Bayh-Dole Act. The act provided an unprecedented deregulation of the intellectual property generated with university research laboratories and facilities by providing for the transfer of the previously automatic intellectual property rights to federally funded academic research to corporations in exchange for exclusive royalty and licensing privileges.

Licensing has been the primary boon for the new academic capitalism in which universities own the property rights to federally funded research which they can and now aggressively license in exchange for often very lucrative royalties. These licensing arrangements also include restrictive, frequently exclusionary proprietary rights on the all research under development at a university site. One of many illustrations is the precedent and case of the University of Utah laboratories which identified and located a gene that plays some significant role in varieties of breast cancer. The university patented and then licensed the exclusive rights to the scientists at their university in an arrangement with Myriad Genetics, despite the $4.6 billion in taxpayer monies that fund research and development (Washburn, 2005).

This large-scale movement to an entrepreneurial university shifts the focus and purpose of scientific research away from a shared "intellectual commons" or community working in and for the public interest. Individual researchers and their teams are increasingly lured or cornered into the contractual and commercial agreements that severely restrict prior publication, dissemination, and the sharing of data, methods, and findings, including any critical and negative reviews, clinical trials, or assessments indicating inefficacy, contraindications, or ecological consequences.

I think it is also worth noting the degree to which academic research is supported by the military through Department of Defense and DARPA-related contracts. State universities, including my own city university, participate in military-related funding programs, including richly endorsed "photonics" research (www.futurepundit.com). "Elite" researchers, especially engineers, are recruited for short-term technological initiatives. DARPA, which produced the original ARPAnet, the progenitor of the Internet, has also been the driving force for the research behind stealth fighters, global positioning satellite systems, and the unmanned drone planes such as those in use in the war effort against the Taliban and Al-Qaida in Pakistan. A former director, Tony Tether formed their Information Awareness Office, which, under Admiral John Poindexter, Ronald Reagan's national security adviser, attempted to create and monitor a national database that would have surveilled the activities of millions of citizens. His successor, Director Regina Dugan, is commissioned to organize contractors, the military, and universities to conduct and produce inventions. In a "balloon hunt" initiative, over 500 teams were brought together to research social networks. *New York Times* reporter John Markoff (2010) writes that the "results suggested the potential of these new ways of gathering intelligence." Dugan stated without irony that she envisions her collaborations with universities as a "renaissance of wonder" (p. 1).

The seismic shift to a capital-intensive, corporative academia prevails in the contemporary climate of deregulation and neoliberal economic policies. Academic disciplines function now under the management of trustees and administrators in a market-driven, bottom-line competitive model in which self-supporting disciplines like the cognitive, neuropsychological, and bioinformatic sciences are deemed exemplary. These fields stand in contrast to the non-lucrative traditional disciplines, including the humanities at large, which require subsidies. Targets for the line of fire of right-wing deregulatory policies manifest in the numerous culture wars of the past two decades; the humanities are losing basic support, funding, and thereby, the very viability of their faculty. They are instances of what Stanley Aronowitz (2007) identifies as the generalized social redirection toward a new "immaterial labor." A new economy of symbol users, of all those that do not actually produce material things or administer people, immaterial labor includes research universities, software workplaces, law firms, architects, and engineers. For Aronowitz in the age of the decline of critical intelligence and the proliferation of technical intelligence "intellectual" in its current connotation, designates immaterial labor, not primarily those engaged in traditional intellectual pursuits such as literature, philosophy and art (doi.www.stanleyaronowitz.org).

The new economy collapses social space, blurring the distinction between the workplace and private space, the "shop floor with the spaces of home." (p. 11).

Habits of Embodiment

The argument over the uses of knowledge carries the echo of ancient and early modernist disputes about knowledge, the subject, and its limits. For the pragmatist tradition, a constant, even vigilant reflection on the continuum of experience is the basis for both education and social development. Resonant with ideas developed later by Freirean-inspired critical pedagogies, epistemology is dethroned from its humanist pedestal. Knowledge in its relations to human interests, power, community, and democracy becomes primary. For Dewey, critical reflection on experience, as the basis for a relational and embodied knowledge is also a personal and collective process of development. Dewey (1929) observed that this process engages "all inquiry and discovery" through a

> surrender of what is possessed, disowning of what supports one in secure ease... For to arrive at new truth and vision is to alter. The old self is put off and the new self is only forming, and the form it finally takes will depend upon the unforeseeable result of an adventure (p. 201).

Dewey once called pragmatism itself a "new name for an old way of thinking." While Dewey regarded curiosity as an ensemble of the human perceptive and cognitive faculties rather than a single mental function, it represented one of the "*luminous points*" of collective agency. It provides a "kindling" to the active and interested development of human relations and experience. In *How We Think,* Dewey (1997) claims that curiosity is the "basic factor in enlargement of experience and therefore a prime ingredient in the germs that are to be developed into reflective thinking" (p. 37).

Mark Johnson artfully reflects on Dewey and William James in articulating a strong sense of the lived and whole experience of "situations." Our perceptions, sense, and pleasure derive from what Dewey (1997) described as an experienced whole. Experience is unified. From the aesthetic perspective

> an experience has a unity that gives it its name, that meal, that storm, that rupture of a friendship. The existence of this unity is constituted by a single quality that pervades the entire experience in spite of the variation of its constituent parts. This unity is neither emotional, practical, nor intellectual, for these terms name distinctions that reflection can make within it (Dewey, 1997, p. 37; Johnson, 2007, p. 74).

The movement of thinking and questioning integrates the passions and the intellectual faculties. Dewey appreciated the workings of this complex, yet holistic unity of perception and cognition, offering an aesthetic pedagogy. It parallels the care for "expression" in all human thought, language, and art that Merleau-Ponty portrays. These convergent appreciations of the situatedness and embodiment of knowledge and perception also help ground a richer sense of the workings of theoretical curiosity

For Mark Johnson (2007), the

> pervasive quality of a situation is not limited merely to sensible perception or motor interactions. Thinking is action, and so "acts of thought" also constitute situations that must have pervasive qualities. It arises from the feeling that a situation is problematic or that it calls out for interpretation and explanation. This initiates a process of intellectual inquiry in search of generalizations that help us understand the phenomena—phenomena that are identified and known only in the context of the inquiry itself (p. 78).

Only in the kinetic interplay of embodied situations does inquiry begin to make objects and distinctions for analysis and explanation. Johnson quotes Dewey again in the context of the pursuit of scientific studies. For Dewey, scientific inquiries "open with the 'Oh' of wonder and terminate with the 'Good' of a rounded-out and organized situation." Johnson is proposing a version of aspects of embodied thought and action first advanced by Dewey, James, and Peirce. I will explore implications of this perspective in the context of a phenomenology that prizes sense from its traditionally diminutive status in relation to the conceptual, epistemic, and cognitive. Johnson's reworking of this premise is timely and opportune for any pedagogy of curiosity. He seeks practices that appreciate and sponsor meaning "grounded in bodily experience." In Johnson's view, meaning

> arises from our feeling of qualities, sensory patterns, movements, changes, and emotional contours. Meaning is not limited only to those bodily engagements, but it always starts with and leads back to them. Meaning depends on our experiencing and assessing the qualities of situations (p. 70).

His work continues a vital, yet minor current that attempts to displace the dualisms that cleave much of contemporary culture in its "education" of the senses.

In James's adventures into a new psychological science, a radical empiricism is set into motion. Thoughts, sensations, and perception are in a continual flux, pulsating to rhythms of embodied minds. As Johnson notes, James's ruling metaphors for thinking are avian, observed, and experienced as "perchings" and "flightings." Thinking is a moving intaglio. There are always at least, latently, impressions within thought. It is not necessarily symmetrical, ordered, and filigreed; certainly, inner speech is associative thought, and, when recounted, its transcriptions appear anything but syntactical, ordered utterances. It is condensed, an amalgam of place markers, minimal and abbreviated mnemonic, symbolic, and body image and sound. Spoken or written inner monologues would lack even a basic grammar. Made intelligible, the associative chain lacks cohesion, is paratactic, loosely jointed, rambling into blind alleys, blurts, pauses, filled with the unsaid in fits and starts.

The traces of the language of thought are symbolically inscribed. Not a pictorial, dream writing, as in Freud, perhaps, but forming a sensory inside, it is a filament that accompanies and coats our intentional acts. Thinking as action is not isolated or independent of the perceptual "buzzing" manifold we can never be entirely detached from without a fundamental existential loss. Johnson illustrates this in James's work, pointing out how even basic grammatical structures are saturated and float in the sensory, perceptual milieu of our preconceptual origins that come to arise into conscious mental acts, concepts, writing, and theory. The abstractions and structures of our speaking and making sense derive from this elemental matrix which is thought's culture. All theoretical forays, even the paradigmatic axioms of mathematics, originate and are capable of continual reinvention, in their return to collective origins where their sparks start.

Thinking thought as action reprises some ancient echoes, including Aristotle's premise that a kinetic aspect is always operating in the natural relations governing animate existence. There is a dynamic reciprocity in the structure and processes of what we now call the "consciousness," "mind" and the body, and all of its attendant querulous study as a basic "problem." The natural philosophy deriving from Aristotle's observations of natural processes applied to the entirety of genetic relations. These came to include the investigation of the changing qualities of any expression of *physis*, as discussed in the first chapter. Inquiry into the nature of living things called upon and questioned doctrines of the intelligible and the nature of knowledge. For some of Aristotle's many heirs, this motivated scrutiny of the development of the physical, emotional qualities of all creatures, including the life of the mind in its pursuit of knowledge. Epistemology in the modern world is one its shrunken heads. From a historical and ontogenetic perspective, thinking, however, is a form of dynamic, kinetic activity, as critically proposed by the pragmatists. This gives a new material voice to Aristotle's tenet that there is a unity to an otherwise double ontology, a "two-fold" being, one that bridges and connects potentials and actuality, knowledge of abstract or timeless, ideal, or cognitive forms with their corresponding experiential and sensuous qualities.

Mark Johnson is eloquent in describing the "flow" of the contact we maintain in our active embodying of unified experience. He portrays the experience of what phenomenologists concern themselves with when they inquire into how the world comes to appear. For Johnson, situated embodiment, inclusive of conceptual and theoretical abstraction, can be gainsaid in part by our attending to how our world comes to "show itself" in the unities of experiential gestalts. He describes spring light near his home in Oregon:

> Toward sunset on what has typically been an overcast spring day, perhaps one punctuated by rain showers, it often happens that the late-day sun breaks through the low clouds and bathes the valley with an indescribable light. Before you perceive *this* or *that* tree, bush, rock, pond, stream, tree trunk, or deer path, you are caught up in the pervading spring-light-bathing-the-valley quality of the *entire situation*. Before you take note of those unique shades of green of the new spring leaves (as opposed to the hard greens of summer, or the tired greens progressing into yellows and browns of early fall), you encounter the whole felt expanse of April greens together. It is out of this pervasive quality of the early-evening situation, here and now, that we *then* begin to discriminate the compressed, intense green of the newly leafed oak from the translucent pale green of the vine maple or the rain-rejuvenated, shiny-tough green of the rhododendron (p. 73).

From the Progressive Era on today, various critically accented pedagogies have found their impetus and motivation in Dewey's challenge to create democratic schooling practices. In all their variety, from pragmatist to critical pedagogical approaches, they argue for a continuum of experience between private and public life and reflection on the practical concerns of the everyday life within a modernity increasingly defined by technology. They converge in linking natural and cultural symbolic expressions and practices and in a refusal of an instrumental and technological reason to schools that structures Fordist and Post-Fordist forms of productivity and scales of efficiency in industry. Critical pedagogies jettison the closed economy of a modernism whose epistemological circuit disconnects historical experience from the production of knowledge, and gendered embodiment from concept and representation. As reviewed in Chapter 2, this is a modernism of empire building and its maintenance. Its instrumental rationality and epistemological zeal is founded on the continual appropriation of labor, both colonial and domestic. Though modernity is a multiple and shifting ensemble of institutions, practices, and modes of social relations both between individuals and the material environment, an instrumental reasoning seeps even inside the dream lives of its inhabitants. A gambling, "inside trader" or calculative way of thinking, instrumental thinking has mutated into a collective mode of not-knowing. Time and space are refigured in a collapse of experiential depth. The new immaterial forms of labor homogenize private and productive spaces. Presence is production. Fermented in the fabrication of humanity's "second nature" within the unchallenged reign of private property relations and commodity exchanges, it is sustained in automatic (and now digital) processes of extracting the surplus value from untold millions contingent, peasant, and outsourced wage laborers across the globe.

Curiosity shares also with intellectualism the stain of some extractable impurity and excess, echoing of its medieval associations with a wandering from essential purpose. Writing in the aftermath of the House Un-American Activities Hearings, Hofstadter observed that "the intellect needs to be understood not as some kind of a claim against the other human excellences for which a fatally high price has to be paid, but rather as a complement to them without which they cannot be fully consummated" (p. 46). As an aspect of intellectual and critical participation into the conditions of experience, curiosity is saturated with cultural affect, and most frequently, deployed by partisan aspiration. As an active social and individual force, it is caught inextricably within a modernity's whose faith in a technological rationality promises unlimited communicative horizons (or their simultaneity in "convergent" media). It always bears significant facets and flaws of character, rivalries, agonism and temporal alliances.

A curious intellect is conventionally personified as either a groundhog or a fox. In ordinary conversation it is often qualified, prefaced by a "just" or "only," that work to keep it contained to a superficial interest. As both excess and lack, curiosity is neither ever sufficiently intensive or "specific" for the development of scientific reason, nor extensive and inclusive enough to constitute a discrete and identifiable discipline of knowledge and practice. Drifting between—even violating the terms and criteria of an efficient utilitarianism—this passion of the mind has little stature in the sober work ethos of Protestantism and the agents of global capital's "flows." Disciplinary studies with their representational languages are often isolated from one another; they are often abstracted from the phenomenal, sensuous worlds of experience from which they derive and might justify their pertinence. Often theoretic, in their divorce from sense knowledge and perceptual dimensions, fields as varied as the philosophy of language or mind, evolutionary and neuropsychology (and in its newer, mainly Anglo-American variants in "situated," "extended," "distributed," and "embodied cognition"), for instance, act to reduce the phenomenal world to terms dictated by their own explanatory protocols, imaging technologies, or experimental procedures. This can act to actually expel the work of sustained inquiry, defying the material making of theory, concepts hypothesis testing, and premises relevant, at least potentially, to sensible human experience and cognition. Academic and intellectual work is justly criticized for its insular enclaves. Provincial in their adherence to canonical "paradigms" or mission statements, they are protected like digital software from becoming infected by the social, political, and rivalry of ethical interests that drive much of our contemporary social practices and research agenda. When they depart from this insularity, as Hofstadter observes (1966), their interventions are frequently challenged as a political partisanship—as illustrated by the fierce and well-funded attacks on climate science—and usually directed toward liberal and "left" advocates of forms of social and environmental amelioration.

A questioning of the everyday, by turns epistemic, "ruthless," or "idle," summons the affective, perceptual, and intellectual momentum to effect qualitative and quantitative changes in the texture and nature of lived experience itself. The social generation and reception of sustained inquiry is inseparable from the various phenomena typically depicted and trivialized as idle or "dispersive" forms of thought and affect. My reading of the notion of critical curiosity is intended to facilitate a reappraisal of the power of this passion of the mind.

Contemporary cultural ambivalence toward intellectual activity persists in idealizing an isolate essence. Developed "northern," states prosper their educational, professional, and intellectual milieu with notions of a disembodied mind abstracted from the textures and rhythms of human social knowledge and its limits. The cult of the genius, the psychic life of the "nursery" (Phillips, 1998) and the "gifted child," as well as the lone pathological actor on the historical stage, remain largely unquestioned. They emerge from a perennial separation of the life of the mind from the events within the layers and connections of our peculiar historical choices and cultural models. Curiosity and an intellectual passion are not preexistent qualities or quantities of the natural world, pure forms of a universal mind or of the uniqueness of human ontogeny. The desire to know is always culturally conditioned.

Social Science as Accomplice

I intend the following as a critical reflection on this pervasive, if enigmatic, cultural force. Curiosity's protean abundance has always been subjected to forms of maintenance, discipline, and control. I wish to explicitly redress the general cultural conflation of a theoretically saturated, critical or "elemental" curiosity in frequent identification as nothing more than an impulse and seeking for novelty, perceptual distraction, or ephemeral attention. I reject any categorical attribution that reifies, idealizes, or trivializes it. It is an elusive energy and force that brings into active perception a common sense that integrates the five senses in active thought and emotion. I do not wish to assert that curiosity is the lone or sole shibboleth and sign for what freedom of thought might mean. It has clearly served the interests of empire building, as well as in the creation of systematic and abstracting, theoreticist models like those of the mathematicians and machinic models discussed in the last chapter. Grounded in premises of a phenomenologically accented reflection, I wish to stress the inherently theoretically informed nature of perception, attention, and sense meanings themselves.

The involvement of behavioral and social science research has been pivotal in an internment of intellectual inquiry. The role of anthropology, sociology, and psychology has both released and disciplined the development of creative modes of thought. In this section, I choose to portray how certain psychological research programs have focused and defined the nature and expression of curiosity. In the following, I wish to survey how this historically entangled passion has been domesticated and neutralized within the disciplinary matrices and power of the social sciences. In particular, psychological investigations into curiosity itself, largely within experimental settings, have assisted the social and cultural legacies of the Enlightenment, simultaneously working toward its naturalization, reduction, and idealization.

Social science research, particularly psychological and psychoanalytic premises are assistants in the modern bureaucratic organization of everyday life. They diligently provide disciplinary categories for institutional practices which sanction some of the inaugural divisions between labor and pleasure, mind and body, and private and public spheres. Epistemic knowledges mark closures within the historically arbitrary representational schema a given social order seals itself inside in constructing its limits to all intelligibility.

I contend that theoretical curiosity is a split image, both idealized, yet suspect, as in the pursuit of "pure science" projects like particle physics or exobiology. It is simultaneously disfigured by pervasive reductive social science methodologies. Protocols premised on the rigorous demand for replicability, verification, and statistical significance, borrowed from the natural sciences, have worked as a means of containment and privation of this primary and orientating aspect of human and animal sensibility. The combined effects of the modernist academic and professional disciplinary knowledges have domesticated inquiry; its affective and intellectual intensities attenuated and channeled into key forms of institutional and conceptual reproduction. The resilience of the desire for knowledge, flowing from its fountains of wonder, ultimately, cannot be ever be entirely tamed or confined by institutional edict or disciplinary canon.

An ongoing entrapment of inquiry is at work in the behavioral sciences. Curiosity is commonly identified as a psychological or "cognitive" drive. It is often regarded to constitute a state of deprivation or essential lack. Psychology and social science research reproduce Protestantism's moral sphere's code of industriousness, utility, and responsibility. It is the domain of desire contained within modernism's physical and psychical economy. By this I refer to an economy of affect and thought caught in the enclosing circuits of repetitive displacements, of desire nursed on lack, deferral, and absence. The electronic compass of digital communication exemplifies this process of depersonalization. Its displacement of bodily motion and exchange is accompanied by displacing language to code. The disembodied "links" within a labyrinthine global circuit disguises the repetition of the same, of an unfulfilled desire. The virtual consumer, weightlessly, shops in the void. A sense of lack, both affective and cognitive, pervades the digital capital mall.

In my own admittedly curious assay of contemporary profiles of this "passion of the mind," I want to explore the formation and articulation of this "cognitive drive." I am committed to displacing curiosity from its comfortable residence in psychological processes, where it is often characterized as a detached or developmentally decontextualized form of naiveté. Social sciences, especially the by-now highly revered clinical researches of experimentalists in neuropsychology and cognitive science, are often profoundly reductive of this complex process involving multiple contexts and relations. Curiosity, as read from a psychosexual and ontogenetic perspective, as reviewed here, requires a complete reconceptualization. A critical curiosity must be disengaged from its historical association with the superficial, absence, or lack.

In departing from a psychological perspective (as framed by institutionalized myopia, narrow technical and formal gauging of context) entails emphasizing curiosity's seminal excess and abundance.

Portrayals of curiosity as a "cognitive drive" antedate contemporary research as well as psychodynamic models. Considered an intrinsic facet of human activity, at least since Aristotle's claim, has sanctified the notion of an intrinsic and ahistorical desire, appetite, and capacity. One question that remains unsettled is the core issue of its origins. An enduring debate continues over the question of whether or not the "appetite" for knowledge is inherent, as a natural drive, or is a process emerging within enculturating frameworks. This has also raised the question of what, if any, constraints exist upon the expression of this drive. If it is naturalized there remains the issue of what constrains or limits are inherent to its functioning and expression. If curiosity is not strictly biologized, as in much of contemporary neuroscience, with its enormous public influence, there is also the concern for the cultural, political, and ideological resistances it confronts. At either extreme, as a naturalized "drive," or as an expression of purely cultural and "personal" interests, there are situations in which it flourishes and is enabled. It is a force within the hybrid of "nature" and "culture" we have always inhabited. Its resistances are channeled into and against the currents of power exerted through institutional and epistemic programs, as evident in the instrumental rationality of Fordist models of production and as they have been vigorously applied to the behaviorism that has pervaded much of the history and practice of American pedagogy.

This conception of desire participates in a larger, generalized intellectual and cultural trajectory, partly Hegelian, and present, for instance, in Lacan's heterodox psychoanalytic premises that place human interiority outside of historical account. Desire is also the animating force, as regarded by Henri Lefebvre and the Situationists discussed in Chapter 4, for ambiguous collective encounters within the closure of the society of the spectacle. The sensuous and the non-sensuous mingle in the strangeness of everyday experience, where things and their representations float between the tangible and the invisible. One antic personification of an all-encompassing concern for dispersal or lack, of a displacement or sublimation, is personified in the pathological obsession about sperm and "precious bodily fluids" in the character of General Jack Ripper in *Dr. Strangelove*. Another figure for setting embodiment and time into a darker vertigo is brilliantly personified by the mad inventor, Jacques Ledoux, in Chris Marker's *La Jette*.

Loewenstein (1994) presents a comprehensive review of American behavioral and cognitive models of curiosity during the twentieth century. He identifies two general waves of research interests. The first mode sought to determine the underlying causes for the search for information, novelty, and irregularity in stimuli presented to subjects. A "second wave," influenced by the establishment of research deriving from the emergence of cognitive and structuralist models in the 1950s and '60s, concerned itself largely with experimental measures for identifying and comparing agreed-upon forms of curious behavior. Lowenstein claims that the search for underlying causes for "first wave" researchers "posed a paradox for early theorists who interpret curiosity as a drive, because drive-based accounts view curiosity as aversive, and hence, seemed to predict that people would want to minimize it rather than seek it out." Lowenstein's own interpretation is that it is a "form of cognitively induced deprivation that results from the perception of a gap in one's knowledge" (p. 76). The issue of drives, discussed below in the context of this overview of psychological research and Freud's comments, persists as a question with significant implications. Experimental, quantitative methods ruled the laboratory roost in the "second wave" of the 1970s, avoiding critical inquiry regarding questions of origins or social and historical variables. Increasingly sophisticated formal research programs developed, replete with their indices, cross-validations, and "curiosity scales." This research tradition, upon careful statistical review, in efforts to correlate scales with behavior, or with individual differences, such as age, gender, and IQ, produced low and contradictory findings.

One of the more prolific experimentalist detectives, D. Berlyne, operationalized curiosity in two dimensions. His "perceptual/epistemic" and "specific/diversive" axes guided many years of research that blended the older behavioral and drive-based models with the newer cognitive and structuralist emphasis in vogue in postwar psychological research. For Berlyne "perceptual curiosity" was defined as a "drive which is aroused by novel stimuli and reduced by continual exposure to these stimuli," with its epistemic other identifiable as "a desire for knowledge" (p. 77). The further distinction between "specific" and "diversive" forms of inquiry attempted to explain differences in attention to content, with "specific curiosity demonstrated by subjects" in their concentration on particulars, that is, puzzles or problem solving strategies, and a "diversive curiosity" manifest by a more general seeking of information, stimulation, sensation, and novelty.

Second-wave studies like those of Kreitler manipulated tasks to seek out conceptual organization, attempting to ascertain the workings of curiosity about complex and ambiguous phenomena. In measures seeking "explanatory behavior" for intelligence, researchers measuring curiosity expected findings to "lead one to anticipate a positive interrelation among curiosity, intelligence and creativity" (p. 79). The tiresome inventory of curiosity scales, including such positivist methodological devices as the Melbourne Curiosity Inventory, which were utilized in pedagogical assessments, presented inconsistent and contradictory findings for a wide range of variables. Langevin's researches, also experimental, claimed to discern differences between subject's curious behaviors in terms of "breadth" as opposed to "depth." Langevin concluded his studies with the open question of whether or not an underlying unity to curiosity, as measured and controlled by complex experimental controls and testing, actually exists. Coie concurred proposing that "perhaps curiosity does not exist as a stable, generalized trait" (p. 79).

Much of this experimental tradition concerned itself also with the enduring and suggestive question of determining curiosity's place within the discipline's ongoing state–trait debates. This involved verifying and assessing the relative dominance of particular contextual situational variations or states in stimulating curious behaviors, as distinguished from a trait as a general capacity or propensity to experience curiosity. Some researchers believed that state curiosity presented a greater research promise than a focus on either diversive or trait curiosity. The distinction is interesting in that it pries open the sealed world of experimental science to issues of social context, milieu, and difference. The trait–state distinction affords inquiry into the roles of contingent and variable environmental resources, cultural and economic capital alike. It opens the laboratory windows to prevalent mores and customs of childrearing and socialization in promoting, harnessing, and hindering curiosity's possible expressions.

Interestingly, it is Freud who first questions the spontaneity of childhood curiosity. In his study of Leonardo da Vinci, he duly notes children's persistent questioning as expressive of a curiosity of circumlocution. For Freud (1964) children at approximately the age of three, this "untiring love of asking questions" manifests the advent of their "infantile sexual researches." Aroused by specific events, such as the birth of a sibling, or another event that appears a threat to their interests, "curiosity of children of this age does not awaken spontaneously" (p. 28). Neither a state nor an inherent trait, it participates within the larger explanatory paradigm of psychosexual drives and their interlocking relationships in repression, inhibition, and sublimation.

The development of curiosity and sexual "researches" follow a parallel course in Freud's formulation. What he calls an "instinct for research" is animated by sexual exploration. The sexual origins of curiosity are Freud's *schaulust*, an ocular lust. This is evident in children's desire to expose and explore their genitalia. This primary ocular compulsion, which he identifies as a "scoptophilia," a "partial" instinct for the good doctor, exemplifies an aspect of transition within the anal-sadistic stage of psychodynamic development. This "looking impulse" impels children to "search for and explore the sexual organs of [his] parents, playmates, and even the inanimate objects of his environment" (Aronoff, 1962, p. 40). In his "Analysis of a phobia of a five-year-old boy" (1953, 10), this looking impulse is made apparent by little Hans's fervid exploration of objects, both animate and inanimate, for the presence or absence of a "widdler." Hans exemplifies a process that Freud considered universal, one manifest in a sexual curiosity that "roused the spirit of inquiry in him and enabled him to arrive at genuine abstract knowledge," learning purportedly to discern the concepts of animate and inanimate objects by scrutiny of a world determinable by the "widdler" criterion (p. 3).

Freud depicts three "vicissitudes" for research in its initial source in sexuality. The first of these is the potential crippling of all exploratory behavior by early repression in which "curiosity remains inhibited for the whole of the subject's lifetime, especially as shortly after this the powerful religious inhibition of thought is brought into play by education" (1964, p. 29). Surpassing the neurotic symptomization of this extreme form of inhibition, Freud identifies a second form in which the instinct for knowledge of the world attempts to overpower and evade the mechanisms of sexual repression. The return of the repressed sexual associations, drives, and inhibited behaviors surface indirectly from the turbulence of the unconscious. In its displacement, it reappears in "compulsive brooding, naturally in a distorted and unfree form, but sufficiently powerful to sexualize thinking itself and to colour intellectual operations with the pleasure and anxiety that belong to sexual processes proper" (p. 30). In this secondary form, investigation and sexual curiosity remain obsessively combined with a growing dependence on "settling things in one's mind" as a substitution for sexual expression and exploration.

The third form in Freud's typology, "the rarest and most perfect," is that of a sublimated sexual drive he finds personified in Leonardo. In this form, the instinctual drives "can operate freely in the service of intellectual interest." Here the labor of libido eludes the snares of infantile repression of the first form, and the compulsive brooding and neurotic deviations of the second by "being sublimated from the very beginning into curiosity and by becoming attached to the powerful instinct for research" (p. 30). Leonardo, whom he dubs the "Italian Faust," personifies this third form in that his life manifested the driving power and passion for knowledge. For Freud, Leonardo "converted his passion into a thirst for knowledge" (p. 24). Even though his life and his unbridled researches give ample evidence of an insatiable curiosity, Freud regards Leonardo's sublimation with ambivalence. Though Leonardo certainly would be an exemplary and robust member of Nietzsche's vaunted breed of new "monsters of curiosity and courage," Freud considered Leonardo's protean and restless intellectual works detour both from his art and from his ability to love. They were diversions of the flow of psychic and libidinal energy, in which "investigation has taken the place of acting and creating as well" (p. 25).

Sublimation, while it frees intellectual powers from repression, also distances thinkers, inventors, and artists like Leonardo from the fulfillment of the drives it counters. For Freud this always involves a certain loss in psychical momentum which is causally likened to finite and limited physical energies and forces. Despite the fact that "he applied himself to investigation with the persistence, constancy and penetration which is derived from passion," including his intellectual works, Leonardo "has investigated instead of loving" (p. 25).

Freud's critical gaze is informed by nineteenth-century principles of mechanical causation, of an energetics, whether hydraulic or physiological, that observes and advances the efficient functioning of working systems and integrated systems of work. Taking the form of psychophysics, the research into sensation and attention by Gustav Fechner, Hermann von Helmoltz, Wilhelm Wundt, Ernst Mach, William James, and others became the basis for the quantification of perception. Jonathan Crary (1999) writes that their research "defined the contours of a general epistemological uncertainty in which perceptual experience had lost the primal guarantees that once upheld its privileged relation to the foundation of knowledge" (p. 12). Freud's interest in curiosity and attention derived largely from the models of Charcot, Binet, and Ribot in its pathological aspects. Attention was central to the experimental sciences. Perhaps in partial response to the intensity of overproduction and the need for increasing surplus value from longer working hours, actual inattention became a problem in the administration of docile bodies. For Crary

> attention is not just one of the many topics examined experimentally by late nineteenth-century psychology but is the fundamental condition of its knowledge. Most areas of research—reaction times, sensory and perceptual sensitivity, mental chronometry, reflex action, conditioned response—all presupposed a subject whose attentiveness was the site of observation, classification, and measurement, and thus the point around which knowledge of many kinds accumulated (pp. 25–26).

From another historically nuanced perspective, Leonardo represents an instance of a current of resistance to the split in experience, both individual and social, between manual and intellectual labor. For Paolo Rossi (1962), the investigations of this polymath demonstrated a resistance to the dominance of scholastic and humanist traditions and their values, evincing the "superiority of the eye to the mind and of the direct and detailed observation of the real world to books and writings." Leonardo's own texts are cryptic, insular, composed in an "obscure symbology and deliberately non-transmittable. Leonardo, who was always curious about a particular problem, actually had no interest in working on a systematic corpus of knowledge" (p. 27). In Rossi's critique, Leonardo's constant and diverse interests into problematics at hand opposed the tenets and techniques of an emerging convergence of empirical and mechanical objectivity of the sciences of his era. An exemplary Orphic persona, he resisted transmitting, validating, or arguing on behalf of his immensely divergent discoveries and interests. Leonardo also caught the critical eye of Paul Valery (1972) who regarded his life and work as illustrative of a problematic in the modernist sensibility in that there is "something that offends in us some of our long-standing distinctions." Leonardo does not conform to modernism's normative expectations, demonstrating the unexpected appearance of a brilliant labor and passion that is "a hybrid, a monster, a centaur or a chimera because of the hybrid species he represents to minds too intent on dividing our natures" (in Posnock, 1991, p. 44).

One effect of drive theories, whether cognitive, behavioral, or biological, is to act as accomplices in naturalizing the desire for knowledge. Both trait and state cognitive models, to varying degrees, share with the very different biological and instinctual formulations of Freud and his heirs in conferring an image of curious work that shrinks it significance to the satiation of appetites and impulse common to all humans across history, culture, and gender. Naturalizing the image of curiosity idealizes the movements of thought into a free-floating, timeless creature of fortune. Popular valorizations of curiosity as a crucial aspect of emergent intelligence and affect, as in Adam Phillips's (1998) quite popular study, *Beast in the Nursery,* are informed by psychological claims that narrowly naturalize the desire for knowledge. An unreflective centering on the particular ethnocentric development of psychoanalytic practice reduces consideration of the multiplicity of the contexts, variations, and experiential resources that provide for its possible expression. Curiosity is reduced to emerging childhood sexual awareness:

> What united, in Freud's view, the artist, the scientist, the lawyer, the teacher, was that they were all interested, in however disguised as form, in the sexual questions of childhood. Psychoanalysis was distinguished by being curious about curiosity, about its provenance and function in a person's life. The child's profundity, in Freud's view, was in the quality of its curiosity (p. 10).

Naturalizing curiosity in the popular imagination accentuates its idle, "innocent" wending within a childhood that does not exist crossculturally. Phillips does provide an ironic reading in finding the interests of the child "as a kind of parody of theoretical or epistemological man, the Freudian child is driven by questions and doesn't believe any of the answers, except his own that he finds satisfying" (p. 11). The rhythms, images, and patterns present to any creative work cannot be reduced into ahistorical accounts of the lives of children. The unmarked child is often the middle-class European exemplar, the inheritor of the objects and toys and novelties. They configure a miniature of the logician or natural scientist, as was the case for Piaget's decades-long documentation of affluent Genevan children, including his own, as epistemic subjects of his relentlessly abstracting "genetic epistemology."

Piaget's enormous influence on the cognitive turn in developmental psychology and generations of educational theory and practice has been maintained at the cost of history. The genetic or "epistemic subject" represented an unbridled will to universalize expressions of intelligence and interaction with the natural world. For Sudhir Kakar (1976), the place of curiosity in Piaget's epistemic project is personified by a changeling, alternately appearing as an expression of spontaneous perception and recognition, and as a conservative seeker of order. Piaget (1951) claimed that "when there is real curiosity, we are no longer in the realm of play but in that of intelligent experimentation" (p. 126). In his tireless documentation of the vicissitudes of exploratory "constructions" of understanding, curiosity makes its appearance in both its perceptive and epistemological guises, modeled on formal logical cognition positions, as a diminutive partner. Though Piaget reworked some of the substance of the underlying principles of causality in his epistemic model, it remained firmly girded by a structuring abstraction. His working model worked deductively, applied to the empirical studies that granted and gave body to his work. The primacy of lived experience is limited to the constructive mechanisms and structures inherent to the development of cognition. It is a minor actor in a tradition, at least since the Enlightenment, in which a formal epistemology dominates over ontology in philosophy, over affect in professional life, and where a disembodied logic remains a signature of Western scientific progress and identity.

Calculating Reason

Mathematical and axiomatic thought processes remain dominant as criteria for the largely quantitative assessment of academic and intellectual performance. In ubiquitous formal and informal regimens of testing, criteria, and normative standards, their influence works powerfully, often invisibly stealing away from social and individual embodied creative, cooperative material processes. Statistical and measurement criteria, as implemented by some of the effects of educational psychology and positivist legions of sociologists and economists, shore up paradigms of a quantitative epistemology and its instrumentation. They inform and are informed by postindustrial models of ideal spatial and temporal production and distribution processes in manifold forms of manufacture, design, and social organization.

In analytic, structural systems, such as the abstract formalism of cognitivist models of mental representation, language and thought apparently have no outside. Their intensive logics claim a fictional transparency, applied readily to tracking, trapping, and referring unambiguously to their objects of inquiry. The cognitive process in Piagetian models participates in a commitment to the same mathematics and abstracting logic of a market economy that creates dematerialized values for the purposes of exchange in a world of commodities. Self-contained and formal, mathematized values preside and rule over the actual material relations and activity of real human agents in social contexts that make them possible.

The impetus generated by a mathematical rationalism has maintained its paradigmatic place, at least since the time of Descartes, as the presiding standard and arbiter of intellectual inquiry. Ernst Cassirrer (1978) claims, "rationalism as a whole has remained firmly convinced that mathematics is the paragon of a priori knowledge and empiricism itself has hardly ventured to touch this conviction" (p. 22). He notes that Kant ventured to critique the ascendant model of the world when he asserted that mathematic is a "shining example of the progress that can be made in knowledge independent of experience" (p. 37).

Serving its own theoretical purposes, mathematics is perhaps exemplary of a purely theoretical and speculative curiosity. It is a self-contained form of thought that permits for the continual development, verification, and examination of propositions and axioms. Its capacious historical projects, which include the geometric theorems and models of Euclid, provided a universal model for the Platonic pursuit of timeless concepts, speculations regarding infinity, recursiveness, Leibniz's articulation of calculus, and Descartes's signal delineation of spatial extension. All of these could provide metaphysical interpretation and precepts regarding the ultimate ground of universal phenomena, appearing to exist outside of lived experience. Plato's theory of ideas "was possible only because Plato continuously had in mind the static shapes discovered by the Greek mathematicians." Classier claims that the "concepts and propositions that Euclid placed at the apex of his system were a prototype and pattern for what Plato called the process of synopsis in idea. What is grasped in such synopsis is not the peculiar, the fortuitous or unstable; it possesses universally necessary and eternal truth" (p. 24).

In an astute historical materialist analysis, Sohn-Rethel (1977) describes the parallel historical relation between the mathematically based "exact sciences" and the rise of capitalism. From a Marxist perspective, mathematically based science acts to disembody the material qualities and relations of physical, productive labor underlying what makes possible all forms of private and social wealth. He claims that "modern bourgeois science is inextricably linked with the capitalist mode of production. It presents a mathematically exact knowledge of nature from sources other than manual labor and other than experience gained from such work" (p. 190).

Mathematical logic and its variant forms of abstracting rationality have held a primary role, both as a paragon of "pure" intellectual inquiry and as a generator of the efficacy of a calculating reason in all fields. Methods of abstraction encoded in commodity exchange relations have provided the model and logical fuel for the machinic and mechanical world of industrial and capital production. Its purported timelessness and unchangeable laws were foundational to the structures and relations that have grounded forms of navigation, weapons, chemical, oil, and mineral manufacture and that yielded the great wealth extracted in the colonial enterprises of the European empires. Abstracting mathematical models informs the structure and genre of new discursive pedagogies of the citizen subjects of the postcolonial projects.

Mathematical models propose implicitly an ontology of number and ratio. Correlations between the existence of things, and of their extensions in space in their calculability within a formal ontology date back to the seventeenth century. Materiality is vaporized in the patterning of a universe in which things and numbers coexist in timeless perfection. Formal and natural science models have flourished within an ethos of instrumentality in which the world of nature is distanced or abandoned in the abstracting processes of prediction, regulation, and verification. Inquiry and discovery produce tentative understandings and relations, formative of the narratives grounding paradigms in the social sciences, often bringing something unprecedented into the world. Formal and calculative rationality adumbrates in the patterning and analytic rigor of the positivisms of the emerging technoscientific schema and the taxonomic, actuarial, and accounting procedures employed increasingly, as Foucault meticulously documented, by medicine, the police, and industrial productive forces. Taxation, census, and security-driven needs like fingerprinting and forms of identity records (now inclosing iris prints and subcutaneous radio- and satellite-guided embedded chips) and their archiving as databases for prediction, control, surveillance, market research, or national security are all debased means of measuring, predicting, and calculating the movements and observable tendencies of select aspects of populations. In the neoliberal, globalized flow of capital their ubiquity has become readily apparent in the control and "accountability" laws, imposing a market share and production-based business model in national testing in the United States.

Abstraction is produced by societies of commodity exchange. It functions in the very production of interest and the inexorable derivation of surplus, or abstracted exchange from actual, conventional use values. It is the invisible signature of the flow of capital. For Sohn-Rethel, "exchange empties time and space of their material contents and gives them contents of purely human significance connected with the social status of people and things" (p. 48). Paralleling Cartesian models of extension, in the abstraction of exchange relations, "time becomes unhistorical time and space ungeographical space; indeed they become abstract time and abstract space, endless time and limitless space" (p. 56).

Natural sciences, in the necessity of their exactness, work to knot together various productive energies to form what Sohn-Rethel terms a "social synthesis." Their abstracting "logic is based on the abstraction from our own timebound existential condition, on the *abstraction of society from itself*" (p. 200). This is not the abstraction of classical *theoria* or speculative, or idle curiosity, but the literally interested, calculating reasoning of modern empires which rose to prominence with a gradual schism set between intellectual and manual labor. In this view, the abstracting processes of the natural or exact sciences and their mathematical models, in conjunction with empirical research, do not derive spontaneously. There are very "timebound causes for timeless logics" (p. 201) gaining supremacy as a historical force to shape the quality of material experience and which now manage the borders of intellectual and creative inquiry.

Psychology and the social sciences are complicit in the institutionalization of abstracting logics and their projects of disembodiment. They cohere in an epistemic regime of stark and austere positivisms that have dried to the bone any intellectual and critical questioning of the "natural" world through their own disciplinary inertia. I find a new complicity evident with the ascendancy of neuropsychology and experimental biological– and genetics-based explanatory models of almost all of human experience and behavior. Both the older and new lions of the "human sciences" have skillfully and artlessly reduced curiosity to empirical epiphenomena subject to laboratory protocols through their own tacit, normative models of behavior. It has been scanned and trivialized to logics of brain, neuronal and biochemical processes. As the generator and fuel for sustained inquiry, active, embodied interest and inquiring remain, however, elements of resistance to any finality, closure, or teleological project. Curiosity insinuates itself as a remainder, the yet to be known or unexplained shifting sand on which the divisions of the social sciences persist. In the strong reading I advance, it is that continuing, experienced moment of thought that poses challenges to the prevailing definitions and conditions of research.

Idealization of curiosity is founded on images of its surprise to thought and conceptual rigor. Its crossings of anticipation and logical sequence are unsettling, troubling to the givenness of things and our relations to them. It becomes a given when naturalized in a popular culture conditioned by largely static social science orthodoxies regarding human nature. As a given of experience, the passion of inquiry comes to be indissolubly associated with the conditions of daily life. Pressing it into the texture of daily life uncritically defeats the autonomy of thought itself. A cognitive divide between perception and thought persists as a conventional wisdom. When considered a natural aspect of childlike discovery within an unproblematic world open to small and unthreatening forays, it is put to practice in forms of social constructivism, process-oriented pedagogies, and textual pedagogies of play or simulacra. All of these allow a liberal polity to find its own reflection. A monadic individual is permitted certain parameters to explore, granting a divisible amount of intellectual property for investigation and investment. Giving it license constitutes accepting the brutal inequalities at the heart of contemporary consumer capitalism, setting off for specialists the principles and adjudications that might be considered for social justice and environmental protection. Curiosity is captivated and captive to the market, retailed, a cherished piece of interior real estate. It is jettisoned from any usefulness; without connection to communities and social movements. It is a legal minor, in fealty to surveilling epistemology and in service to endless productivity.

In Freudian and cognitive models, curiosity as a naturalized given, necessarily entails the pursuit of the means of its own satiation. It is likened to all other appetites and thus is constituted by a primary absence of its object of gratification. The notions of "gap," "lack," or "lapse" call for a historical interpretation and appropriation of interests. How did these notions become entrenched within both social science and the popular imagination? The question of desire arises like the sexual appetite. A questioning of desire is also the desire of a questioning subject. It clearly reflects the modernist anxiety about the fulfillment of basic needs and the estrangement of desire in a technologized metropolitan world. It also opens the question of the durability and ubiquity of privileged modes of life. How time bound are the sustaining meanings, systems of belief, habits of embodiment of "Western" or "Northern" dominant cultural models of culture and community, including its scientific cultures? Obviously, there are multiple, vying models of practice, coexisting to varying degrees of acrimony and cooperation. A closed economy of desire upholds an inherent and abiding absence or lack in human existence. It is an economy founded on theological or secular proscriptions and injunctions; it is a world dictated by law and dominated by necessity, fatalism, and determinism. Social conditions, including disease, hunger, poverty, and prejudice are determined to exist somehow beyond and outside of history or human agency. It is a shrunken, desiccated earth, one on which freedom to act is predetermined and constrained.

Historical subjects themselves become the curious objects of knowledge. In the contemporary context of technoscientific management of life through genetic engineering, as discussed in Chapter 6, there has been an intensification of a relentless questioning and objectification of the human species and its limits. The knowledges deployed to shape, control, and, sometimes, release the bodies of the medical, psychological, penal, and pedagogic subject have developed unique techniques and methods. Biopower, or the regulation and management of populations, combines theoretical and technical curiosities in the interest of regulating, legitimating, and giving coherence to normative values of sovereign states.

Nietzsche (1966) offers a description of this impulse in *Beyond Good and Evil.*

> ...obedience over a long period of time and in a single direction: given that, something always develops, and has developed, for whose sake it is worth while to live on earth...the long unfreedom of the spirit, the mistrustful constraint in the communicability of thoughts, the discipline thinkers imposed on themselves to think within the directions laid down by a church or court, or under Aristotelian presuppositions...all this, however forced, capricious, hard, gruesome and anti-rational, has shown itself to be the means through which the European spirit has been trained to strength, *ruthless curiosity*, and subtle mobility, though admittedly in the process an irreplaceable amount of strength and spirit had to be crushed, stifled, and ruined (section 188, p. 101).

A "ruthless curiosity" is the realm of the specified, of ontological difference contained. It brokers predetermined categories of existence: a fateful presence, in a remaining of the same and a resistance or inviolability to difference, transformation, and multiplicity. Difference is a stigma or marker: it is an object of seizure and appropriation and exhibition in the *wunderkammer* of calculability and control. Difference is a primordial force within the unity of things. It may compel the pursuit for secure ground in explanatory, epistemological foundations or in interpretive matrix theories, like those offered by complexity and "emergence" studies, which provide moving shelters for convergent phenomena. Difference gives voice to an enduring nostalgia for a unity of thought and being. The Promethean persona is invoked as the master who would surpass all existing limits, mysteries, and problems to synthesize knowledge into a final totality. Master discourses, like their theological cousins, marshal their forces in the arrest of unruly becoming and errant difference. This "ruthless" assertion of an imminent explanatory model is evident in some the of research programs today, like those conducted in neuropsychology laboratories at

MIT and Carnegie Mellon, in the boundless syntheses proposed by Stuart Kaufmann and Gell-Mann at the Santa Fe Institute. Whether at the genetic, or nanoscale or cosmological levels of inquiry, the former governed by dogma of the nucleus of genes, the latter by Big Bang, supersymmetry, and unified theories, these ongoing projects proffer a high degree of certitude, explanatory symmetry, and design for the disorder, incongruity, and apparent randomness of empirical phenomena.

Critical Interventions

Audibly accenting a distinctively epistemological curiosity are pedagogies engaged with the problems and potentials of daily life. An epistemological curiosity, in the language of critical pedagogy, as articulated by Paulo Freire (1998), animates the social production of new knowledges. Forms of new knowledge emerge from a primary and "ingenuous curiosity," an inherent human interest to question, wonder, and worry things. An "ingenuous curiosity" is the daily engagement of the mind in the world, thought's place, its bodies and the substance of material sensuous experience (p. 35). The ambiguities and contradictions of quotidian life are made into the object of a critical gaze. They are continually problematized and reworked in the collective praxis of pedagogy. Reflected, questioned, and purposefully placed in dialogic relation and historical context, an epistemological curiosity gives contour to common sense knowledges and relations to the world and its remaking. Fashioned into the tools of thought, constantly sharpened, and given edge, critical pedagogies, deriving from the Freirean "culture circles" in developing countries, sustain pedagogical interest in how the world is given sense. Its pedagogical objects vary within regional and local contexts and problems of existence, but the Freirean-inspired motivation is premised on a critically informed epistemological curiosity.

As a goad to creative activity, curiosity is a key thread tying learners and teachers together in collective reflection and action in daily life. Exemplifying what I am identifying as an abiding critical curiosity in attending to the manufacture of social processes in general, it is a practice especially inquisitive about the processes and relations of human work and the production of knowledge, including its own. As Freire (1985) suggests "one who studies should never stop being curious about other people and reality. There are those who try to find answers and those who keep on searching" (p. 2). Critical curiosity is a name for the experience of futurity—as an opening without determination, a becoming different, resisting representation and fixed subjectivity. In the work of Freire, it is also the experientially driven basis for any literate knowledge; primary to all learning "curiosity is what makes me question, know, act, ask again, recognize" (1998, p. 81). A cultivation of the sensuous and sensibility, critical pedagogies claim a constant engagement with the needs and aspirations of a people in a given time and place.

In its concern for the development of agents and producers of knowledge, Freirean critical pedagogy places its accent on surpassing the imposed limits of codified knowledge and conditioning. In stressing the primacy of critical reflection on experience, it parallels the thought and practice of Deweyan pedagogy. It is steeped in a phenomenological appreciation of the experiences of individuals and collectives. It draws from the deep well of social experience as the source for description and critique of shared, if incomplete, knowledges and relations. Freire also wishes to convey the rigor and persistence of a pedagogy of curiosity in claiming that the necessary conditions for critical learning

> demand the presence of teaching and learning simultaneously in the context of a rigorous methodological curiosity anxious to explore the limits of creativity, persistent in the search, and courageously humble in the adventure. In these conditions, those who are engaged in critical learning know that their teachers are continuously in the process of acquiring new knowledge and that this new knowledge cannot simply be transferred to them, as learners. At the same time, in the context of true learning, the learners will be engaged in a continuous transformation through which they become authentic subjects in the construction and reconstruction of what is being taught (1998, p. 33).

Freire's legacy impels a questioning of the constitution of mundane life, and the conditioning and socialization of historical agents. In pedagogic and political interventions extending across the hemispheres, but primarily among rural, underdeveloped communities, Freirean practices manifest a coming to question how primary and basic conditions of existence and culture form and develop. It is a restless practice, constantly inquiring into the organization of daily labor, tasks, and the distribution of roles and responsibilities between individuals, castes, classes, and gender. The culture circles favored by Freirean-derived practices challenge the distribution of privileges and resources, both economic and cultural. As an epistemological curiosity, students, often rural agricultural workers, like those in Recife working in the *latifundia*, ask how their lives came to be impoverished, ask why there is poverty and wealth in the same time and place, investigate the origins of disease and danger, and rethink the development and qualities of natural and social relations. Students question how other people live, researching the primary question of whether a better life than the ones their communities have always lived is possible. As examples of an epistemological and social curiosity not separated from practical action to change the conditions of lives, they are informed by the energies summoned from a deeply passionate interest, engagement, and reflective connection to routines, rituals, patterns, and practices of everyday life.

Freire's commitment to a dialogic and liberatory pedagogy cannot be taken without scrutiny of the epistemological and critical curiosity he invokes. The Freirean project was premised on an autonomous coming to knowledge through critical engagement with everyday life. It is premised on the social development of knowledge in which the objects of experience and understanding are formulated through continual questioning, use, and reflection. The dialectics of knowledge are communally enabled through the culture circles and educational settings acting as experimenters in a social laboratory. Freire invokes the expression and use of a "creative" and "epistemological curiosity" through which a full comprehension of the purposes and history of the objects of knowledge—especially of the dynamics of social relations between classes—comes to fruition. The process of "conscientization" that enables the transformations from an "ingenious," common sense, toward an "epistemological" curiosity requires a rigorous and unrelenting willingness to reflect upon daily life and experience. Reflection on motivation, attitudes, and social conventions, as well as the use of common objects in daily life, like the availability and access to potable water, medicine, or textbooks, form the basis for the critical examination of experience. The pedagogical development from common sense to critical reflection, for Freire, creates educational milieu that "maintain alive the flame of resistance that sharpens their curiosity and stimulates their capacity for risk, for adventure" (1998, p. 32).

Advocating and presuming no absolute origin, process, or medium, a critically interested pedagogy is an unpredictable activity. Eschewing questions stemming from natural scientific method and their ahistorical criteria of truthfulness and validity, a pedagogy of curiosity sponsors the emerging possibilities generated from a constant questioning of experience and culture. A curriculum of experience, as in certain pragmatist renderings discussed here, is a pedagogy of encounter, to act, know, and test the limits of existing knowledge—relations, against which daily life is constructed and maintained.

The changes brought about by thinking through the possibilities at hand, in light of collective, accumulated practices and knowledges are expressions of social and critical curiosity. They are the compulsions, directives, and desires through which individual or collective subjects are likely to crystallize an acuity directed toward the historical realization of material and symbolic resources and potentials. It is the energy that ignites thought; like the swerve and swirling motion of atoms in Epicurus, sparks of interest, intent, trouble limits to resist the constituted order of disciplined desire. As thought's own swerving, curiosity becomes manifest in ways of reading and writing selves, researching, organizing communities in struggles to attain the knowledges requisite to understand the historical formation of injustices and their potential resolutions.

John Elias (1994) points to an ambiguity or limit to a notion of a "conscientizing" educational practice that does not fully take into account the degree to which it compels or manipulates its idealized subjects. The expectation that certain critical insights and positions will unfold through the question posing process could limit the kinds of expression of intelligence, curiosity, and engagement with the world. For Elias

> it is Freire's philosophical position that an objective reality exists, which all will inevitably come to recognize through education. This almost quixotic view fails, however, to do justice to the complex nature of reality and of human knowledge of it. It leaves little room for relativism and a pluralism of world views (p. 104).

Elias is taking to task a strictly realist account of objective knowledge. Dogmatic and reductive variants of historical materialism and critical pedagogy are grounded in claims to this kind of objective knowledge. It raises the contradiction of imposing a pedagogic system for the freeing of thought for social action. The Freirean legacy, despite its invocation of a language of realism, or an adaptation of reflection theory between the world of objects and knowledge, is too heterogeneous to be taken to task for this reduction to an abstracting objectivity. Freire's eclectic philosophic sources are invigorated by the descriptive richness of a phenomenological approach to describing, rather than providing a priori principles or explanatory frameworks for the problematics of lived experience. In addition, the original tenets of critical pedagogy were formulated by an unprecedented mixing of theoretical practices, which, in addition to phenomenology, included liberation theology as well as a reworking of Hegelian-Marxism in the context of the Americas. It provides the basis for generating situated critical pedagogies throughout the world that emerge and remain ultimately grounded in the contradictions and ambiguities of lived, sensuous activity. At stake in Elias's criticism here is a contemporary charge in the endless trial of curiosity itself: the descent into dogma, method, and doctrinaire practices and the need for a restless resistance to formulaic responses and habits of thought, even, and perhaps especially, for those that would claim to work toward ways of reknowing and rewriting the world for freedom and social justice.

Given a new image to the experiments and encounters of thought, contemporary pedagogies must accentuate its particular and peculiar eruptions, rhythms, and motions as they are given form and contour by embodied historical agents in their ongoing questioning and reflection on the problems and possibilities of actual experience. A pedagogy grounded in the uncertainty of curious thought will explore and be propelled by its irregular but constant movements, its hesitations and leaps, tracking the vicinities of thought, passion, and commitments it comes to inhabit. Their curriculum and methods are generated by key problems and continual questions from multiple perspectives and traditions. It is a pedagogy that would take upon itself the task of reconceiving human agency in relation to the natural world and the disconnections that have caused misery for most living beings across history. It is an attempt to reformulate the purposes and directions of education as a socializing process. It is nothing less than the project of reimagining the relations of knowledge. Critical pedagogies parse the spectrum of embodied, gendered, and raced experience for material connections in their actuality and virtuality alike.

The curious scholar I wish to figure here "responds to what in life as in thought is not already determined—to those unforeseen moments in what happens in us and to us that open up onto new histories in history, new paths in the 'complication' of our ways of being" (Rajchman, 2000, p. 61). This is the student of singularities, one who discerns new linkages in the contingent, shifting circuits of thought, agency, and objects. Crossing conceptual and disciplinary boundaries, pedagogies of the question pursue particular problems and disclose differences and singularities pertinent to lives lived under specific conditions of possibility and constraint. Investigating relational fields, juxtaposing and partaking of a certain intellectual violence to the integrity of closed systems and finality, they are foragers of possible knowledges and its multiplication in new, unforeseen networks, contacts, and alliances.

Critical curiosity is manifest in acts of intervention into daily life in the interest of transforming it. It is never entirely independent—even in its purportedly transcendent, speculative aspects and personae, it is motivated by specific historical events and their contours in thought. Epistemic, critical intellectual work operates within the quotidian conjunctures of alliances, technical instruments and forms, juridical and political policies, and funding and investment strategies of commodity culture. Intellectual practices and the conditions of possibility that open opportunity for a critical curiosity to permeate activity, operate in specific relations of power and resistance. The fragility of communities throughout the world, of refugees of war and "failed states," of those millions of dispersed and immisserated groups give substance to the notions presented by Giorgio Agamben (1993) of all those with no consistent, viable, or necessary identities. The situations that arise activate potential responses to contingent moments, to unexpected possibilities for intervention. These confrontations with contingent situations arise continually in the inherent crises of a world of finance capital and management of the earth's resources. In their ubiquity and often calamitous impacts they are also apertures for the creation of future communities. Come brazenly to the fore in their brutality—as in the mass rapes attendant upon the wars in Kosovo, Sierra Leone, Chechnya, or the Congo or the conditions in which women live under Taliban control—they offer instances in which reshaping daily life is certainly more than called for. Thinking back to the last decades, there are a plethora of contingent crises that ask of us how and when will there be the forces to create alternative communities, ways of cohabitation not premised on national, ethnic, or raced essences. Among the many that come to mind are the legacy of Reaganomics, the HMO model of privatization of health care and schooling in the United States; of eruptions like the disintegration in the 1990s of Yugoslavia, the Hutu-Tutsi genocide in Rwanda and Burundi, the civil war in Sierra Leone with its child soldiers; of the sudden composition of a global and environmental justice movement in the "post–September 11" milieu; of a continuing AIDS-HIV crisis, especially in sub-Sahara Africa; and the ravages experienced by the Iraqi populace despite international condemnation of the Iraq war as an unnecessary war. Persistent structural issues like immigration, racism, health care, accessibility to drug and vaccine treatments, and reproductive technologies clearly mark central concerns for any genuinely critical pedagogy and for all intellectual and critical interests today.

While John Michael (2000) argues for a reclaiming of the Enlightenment's universal aspirations in arguing for democratic pluralism and social justice—one requiring taking to task the postmodern disdain for transcendent values and principles—critical curiosity can only find expression through moments and situations resisting final principles and calling into question existing dogma, beliefs, and social dynamics. These have included confronting ethnocentric, classed, and sexist notions of a universal "human nature." The new essentialisms have reared their ungainly heads in the guise of sophisticated genetic and neurological models of human behavior. The entire spectrum of human activities, including affect, motor, motivational, and "higher functions" such as creativity, memory, and aesthetic sense, are now represented and archived in fMRI (functional magnetic resonance imaging) technologies. While there is value in their contribution to diagnostics and potentially early interventions and treatment, these models collectively serve to naturalize human activity and potentiality. They are second- and third-order representational forms. Their connection to actual human behavior and sense of well-being is tenuous, if not absent. Neurobiological models, like those that had originally formed the hyperbolic health claims for the multibillion dollar investment in the Human Genome Project, present biological, ahistorical, and often determinist explanatory frameworks. Outside of a "history of prior conditions" for individuals, a sense of agency and the role a state and society plays in guaranteeing the welfare of its citizens, remains offscreen.

The affirmation of embodied experience is the point of origin for a pedagogy of critical inquiry. It is a practice accenting the differences and commonalities of shared communal life. Despite its perennial usurpations, it is an active and reactive element of human experience; it is corporeal, perennial, and insuppressible. Knowledge's bodily reserves are incorporated, as Foucault (1970) asserts, through actual lived experience.

> Actual experience is, in fact, both the space in which all empirical contents are given to experience and the original form that makes them possible in general and designates their primary roots; it does indeed provide a means of communication between the space of the body and the time of culture, between the determinations of nature and the weight of history (p. 321).

To consider the everyday in a critical perspective opens the possibility of projects and perspectives whose concern is the quality of lived experience, knowledge, and the making of the future. Pedagogies of curiosity either set into motion or deter programs, organizing efforts and agenda that would seize historical contingency as a starting point for reevaluating existing modes of thinking and acting. Any pedagogy starting with historical and daily experience can begin the process of creating cultures undivided between a theoretical or epistemological and a "dispersive," or "idle" curiosity. A pedagogy founded on a critical curiosity refuses to accept the foundational binaries fabricated on pivotal historical fault lines in Western culture and thought, including those long established between intellectual and manual labor, between knowledge reflective of embodied, material experience, and the abstract, formal model of mathematics invoking a universal, transhistorical mind.

A critical curiosity is only possible as an expression of a developing critical community. A critical community is the ongoing process at work and in the pleasure of a community of questioners, what Pierre Bourdieu (1999) called for in the development of a "collective intellectual." It is in the spirit of developing collective initiatives, such as those explored in the previous chapter, when Foucault (1996) states:

> Curiosity is a vice that has been stigmatized in turn by Christianity, by philosophy, and even by a certain conception of science. Curiosity, futility. The word, however, pleases me. To me it suggests something altogether different: it evokes "concern": it evokes the care one takes for what exists and could exist; a readiness to find strange and singular what surrounds us; a certain relentlessness to break up our familiarities and to regard otherwise the same things; a fervor to grasp what is happening and what passes; a casualness in regard to the traditional hierarchies of the important and essential.
>
> I dream of a new age of curiosity. We have the technical means for it; the desire is there; the things to be known are infinite; the people who can employ themselves at this task exist. Why do we suffer? From too little: from channels that are too narrow, skimpy, quasi-monopolistic, insufficient. There is no point in adopting a protectionist attitude, to prevent "bad" information from invading and suffocating the "good." Rather, we must multiply the paths and the possibility of comings and goings (p. 19).

This community will always be contingent upon those who share a disposition to act in concert in their power to pursue the uncertainties, doubts, and dissonances in collective life. For Bourdieu (1999), it is the work of this collectivity to participate in the creation of "the social conditions for the collective production of realist utopias" (p. 19). It is a community of those who wish to embody ideas in activity, which are willing to trouble the false symmetries of both bourgeois and dogmatic revolutionary thought alike. Critical communities have arisen in movements like the groundswell of projects leading to the divestment struggles against apartheid in the 1980s, the anti-sweatshop labor practices in the '90s, and the agitation and dissemination that has challenged financial austerity arrangements like the World Bank and the WTO, as well as the communities of activists concerned with genetic diversity and bioengineering. Heirs to the Situationists perhaps, discussed in the next chapter, these recent and unexpected upsurges manifest resistances that recurrently spring up in response to hegemonic practices attempting the global administration of everyday life.

A "questioning of the everyday" as a social curiosity, by turns epistemic, ruthless, or idle, summons the momentum and power to effect changes in the texture of embodied being and becoming. The social nature of curiosity, as a passion, tamed and resistant, disciplined and abundant, is inseparable from the phenomena typically trivialized, even vilified, as "idle," perceptual, "dispersive," or ephemeral, often feminized, forms of inquiry. My reading of the notion of critical curiosity is intended to be felicitous in encouraging a core reappraisal of this abiding desire. This reappraisal is necessary as a contributing force in not only critical pedagogical practices, but more originally as an element in the fusion and taking shape of all conceptual, theoretical work and the innovations they generate. It is a vital aspect of what phenomenology (as discussed in Chapter 5) identifies as the intentionality of all acts of thought. A pervasive psychologism in reductive social science methodologies has disfigured this protean power, confining its role to the interiority of isolated subjects. These protocols have worked as a means of containment and privation for the disorderly, self-serving, and unsystematic "zeal for knowledge," as Augustine, with considerable virulence and ambivalence, identified it. The combined effects of the disciplinary knowledges have been both an immeasurable restraint and domestication of inquiry; its affective and intellectual intensities attenuated and channeled into forms of production.

In order to place an emphatic stress on its social volatility and mutability, I have not intended to portray it as an avatar of some innate vitalism at work in the dynamics of social production. To name just a few contemporary currents of inquiry—bioengineering, critical pedagogy, particle physics, exobiology, cognitive science, and cultural studies—are all coalescences, assemblages, and alliances of interests and passions. They are practices generated by and through alternating and mixed energies. They give expression to their composition as combinatory social tendencies by elements or alloys of ruthless and epistemological curiosities—at once kinetic, entropic, and catalytic. In Deleuze's (1983) use of Nietzsche's idiom, they are polyvalent wills to power composed at the same time by active and reactive forces, in shifting intensities and alignments, patterns, and systems of thought in motion, relation, and space.

Inquiry and discovery produce tentative understandings and relations, formative of the narratives grounding paradigms in the social sciences that often bring something new and unprecedented into the world. Curiosity is never an essence, mental state or stable, identifiable concept. The perennially contested and shifting historical positioning of knowledge as social power and passion requires an appreciation of curiosity as an emblem, decoy, or secret agent of material interests and domination. The duality of a purportedly disinterested knowledge, of a contemplative life enabled by the idle pursuit of the purely theoretical and speculative, as contrasted to its debased mechanical and manual counterparts serving utilitarian ends, was and remains a self-serving mythos.

The place of technics, discussed in Chapters 5 and 6, aids in considering the partitioning of knowledge and practice. In regarding the technoscientific discourses and practices of bioengineering, genomics, artificial life, and exobiology and extraterrestrial studies, I attempt to profile some of the contemporary resonances of curiosity's contested place in social relations, the public sphere, and the powers of thought in and about the nature of everyday life.

To be possessed, curiosity—and the generative experiences in which it thrives and sparks—is kidnapped, pirated, and put into conceptual service. As a conscript, curiosity yields to thought's labor, willed to the abstracting production of concepts, representations, and essentialized identities. It is never any purer than the truths that are its provisional patrons. There are no essentially "ruthless," liberatory, or "epistemological" and disinterested assemblages of relation and difference that would give curiosity a fully autonomous, pure function and value. Curiosity is never an essence, mental state or identifiable concept. I contend it is thought's freedom.

Chapter 4
The Sphinx

> *He turned into Cumberland Street and, going on some paces, halted in the lee of the station wall. No-one. Meade's timberyard. Piled balks. Ruins and tenements. With careful tread he passed over a hopscotch court with its forgotten pickeystone. Not a sinner. Near the timberyard a squatted child at marbles, alone, shooting the taw with a cunnythumb. A wise tabby, a blinking sphinx, watched from her warm sill (Joyce, 1990, p. 77).*

The everyday remains largely undertheorized as the locus of individual and collective transformation. The world of daily experience is the crossing point and frontier for critical interest, theory, and passion. The quotidian, comprised of habitual contacts, frames of reference, identity formation, and solidarity, presents the matrix of experience, providing the context and content for the workings of a critical theoretical curiosity. As I will discuss in this chapter and in the phenomenological accent of Chapter 5, the everyday presents the vital sources for interpreting the convergences of affect and theoretical reason. It is the locus for a reassembling the losses of communal and historical experience within contradictory forms of contemporary social organization, temporality, and space. It is the experiential crossroads at whose junctions philosophy makes its forays as a critique of existing social space, conditions, and problems. As a mobile site for a "history of the present," the everyday announces a threshold for critical theoretical projects.

In this chapter, I will describe specific intersections of a critical, theoretical curiosity in postwar modernity. My intention is to portray their transversals as critiques of everyday life, conducting currents of theoretical curiosity as practices that acknowledge the contingency and aleatory aspects of lived experience, spurring a continual reinvention of intellectual and critical thought. In their departures from a scientific model of dialectical materialism, thinkers like Lefebvre, Debord, and Althusser, an emphatic interest and care characterize the workings of critical thought. Their critical engagement is propelled by an intellectual passion

An earlier version of this chapter appeared in *Situations*: *A Journal of the Radical Imagination* (2005), 1:1, pp. 44–74

M. Zuss, *The Practice of Theoretical Curiosity*, Explorations of Educational Purpose 20, DOI 10.1007/978-94-007-2117-3_4,

for not only knowing and discerning the intricate mediations within and between present relations and conditions, but for seizing opportunities for their structural transformation. The elusive, often oblique powers of theoretical curiosity stir the green fires of what Hume (1965) called one of the "calm passions." In the interest of subverting the mundane, this "passion" of the mind interposes itself between the cracks of conceptual order and rationality. As both excess and trace, it plies through the fibers of ossified historical and political organizations of representation, to counter the cohesion and logic of the proper, the placeholdings of language and a frozen symbolic order.

Critical theoretical curiosity is a form of disquiet, an abiding and resilient doubting or troubling of the fixed and familiar. For writers like Lefebvre and Althusser, it is an ongoing intellectual operation, a willingness to test and try the prevailing tensions and antagonisms of social life. It maneuvers in tactics of waiting, detour, and intervening when opportunities arise or, in provoking activity, keeping all premises, including its own theoretical development, unsettled. It is a detective work in vigilant scrutiny for novel patterns and resistances arising within the arbitrary, repressive, and discontinuous order of social relations. In its variant formulations, theoretical curiosity is generative of counter-constitutive forms of power. It is sustained and given momentum within the materialist encounter and contingent convergence of thought within the everyday.

Theoretical curiosity is always situated in alliances and strategic interests and positions within a continuing class war within theory itself. This was illustrated by the very distinctive and antagonistic formulations developed between the humanist practice of Hegelian Marxists like Lefebvre and the theoretical anti-humanism of Althusser's circle. The debates over questions of humanist practice also entailed foundational questions regarding the limits of representation and historical agency within Marxist theory and practice. For the postwar critical interventions reviewed here, a critical curiosity always involved participation and experimentation. They are vivid instances of what Merleau-Ponty called "adventures of the dialectic," of the power and labor of the negative. In their attempts to fuse theory and action and experimental praxis, they engaged a panoply of aesthetic and intellectual activity for a rethinking of the nature of human as a political animal. Not only of historical interest for critical pedagogues and activists in social movements now, their interventions signaled potentials and strategies for rewiring circuits of cultural and economic power. Their tentative, ambiguous legacy still provides suggestions for ways to regenerate the totality of forms of knowledge and social life.

Theoretical curiosity conventionally connotes inquiries that test and resist imposed limits to thought within existing forms of intelligibility. The forms and practices we inherit and that form the representational basis for the "natural attitude" or habitus emerge from the enclosure of experience, perception, and knowledge with a given historical episteme. This is illustrated by the heterogeneous phenomena which have variously been identified as post-Fordism, "empire," globalized, "late," or network, capitalist postmodernity. Theoretical curiosity presents a cloud chamber in which the traces and collisions of thought can be observed as they radiate from disciplined epistemological centers of gravity. In this chapter I want to advance the claim that theoretical curiosity is expressive of the configured mediations of sustained practices of thought in their emergence and resistance within the incalculable productivity of any social ontology.

The writers discussed here evince a theoretical curiosity and care for the unknown valences and resonances of the phenomenally contingent. They are multifaceted, heterogeneous practices setting into motion oppositional tenets to all expressions of totalizing and abstracting rationality. They act as calculating, experimenting exponents for the aleatory or contingent. The heterodox tenets of these postwar thinkers are divergent in important ways, yet they cohere in a rethinking of the negative and contingent, as material appropriations of convergences of theoretical critique and historical experience. They generate particular ways of theorizing upon what for historical materialists before the war had remained the scantly theorized issues of arising from consideration of the everyday. In resisting their own critical resistances, they rarely make a complete rupture with the forms of rationality they would subvert. Whether capitalist exchange value relations, French Communist Party and reductive economic Marxism, or the ghostly and restless negativity of Hegelian dialectic, they continued to resist in both senses, propagating a theoretical curiosity engaged in everyday life in its plenitude and contingencies. Althusser and Foucault, in particular, suggest cunning interceptions into the managing operations of what otherwise appeared the unsurpassable totality. They confronted the constituent discursive powers at work in forming subjects, political parties, and states, as well as in the abstracting logic of a social synthesis that survives spectrally through the commodification of life.

As advanced by critical theorists of everyday life, contingency provides neither a mere potentiality, nor a calculability and cunning of reason. It exists within a plurality of intersecting, aleatory worlds of the virtual spheres of knowledge, relation, and experience. They need not, out of necessity, become manifest. Theoretical practice is a material force within cultural history. Contingency is the hesitating beauty of history. It names a subjectless coincidence of material forces, dispositions, and apparatuses, at once physical, conceptual, and identifiable within social histories of struggle. Projects generating the energies deriving from the aleatory and contingent ply the unforeseeable depths of actual historical conditions of possibility. Their practices run the often-random course, momentum, and current of histories of the multiple and convergent, of "now-time," of Walter Benjamin's *jetzeit*. The contingency of praxis is never accidental or capricious; curiosity's singularities are not pure chance encounters even while they resist systematization. A resistant, insurrectionary curiosity hovers cunningly, prepared to pounce across the borders of constrained time and space of the everyday. It seizes time. It detects the point and place of a situation for the release of collective desire. The critical interventions discussed here ready their practices and tactics for integration into the irreducible momentum of bodies and minds moving beyond the limits of what Marx (1992) observed as a "civilization trapped within the crude barbarism of need" (p. 363).

Theoretical curiosity as a practice is distinguished from its conceptual counterparts, especially defamiliarization, estrangement, and the uncanny, in the workings of thought that it conducts. In thinking the singular, theoretical curiosity confronts its own premises, working thought's historically imposed constraints, including the production of the givenness of its theoretical objects. Sharing with practices of defamiliarization, a decontextualization of unreflected objects of experience and perception, theoretical curiosity encounters the contingent. It wields its formative powers as it distances itself from elements of its own premises and arguments; as a consequence, it is dialectically changed within those encounters. As a materialist practice it is engendered and embodied in social relations and the antagonism of everyday life. It is not disinterested speculative practice, expressive of the leisure and remove of classical *theoria*, the realm of Hegel's "spiritual animals," who, in Pierre Macherey's (1998) phrase "remain without works," never partisans, they remain "devoted to the pure practice of [his] capacities" (p. 31). A critical theoretical curiosity, *pace* Spinoza, refers here to the power of thinking-bodies, at the individual and collective level, to act in sympathy with each other in their natural right, in their disposition and powers to coalesce and cohere as thinking and acting interrogators of the dominant, if discontinuous, epistemic structure and coding of social experience. As reciprocal and mutually interacting bodies, the passion of a critical curiosity is generated and augmented by collective social movements, interests, and interventions that engage and challenge the nature and constitution of the everyday life. It is a primary catalytic element for thought's pursuit and production of an open-ended futurity.

Theoretical curiosity as a practice operates and is generated by the specific historical experiences and instances of contradiction. In the interest of mobilizing transformation, it acts within and against the grain of everydayness, a friction within the ontic. Arising from the conditions of possibility, as much actively present as absent from a historical conjecture, theoretical curiosity is alert for unexpected opportunities and moments, readied for the aleatory gambits that might suddenly be offered. It is a critically informed practice, one that confronts philosophical and disciplinary limits in thinking philosophy and theory's own outside. Theoretical curiosity problematizes the immediate and generates the labor of Foucault's "specific intellectual." Foucault, who, as I cited in the last chapter and who "dreamed of a new age of curiosity," considered his genealogical studies and "history of the present" as a process of constant problematization. As a philosophical immersion within and against the facticity of the everyday, it proposes movement from stasis and reified determination to a collective and permanent process of multiplying ways of living and knowing.

In Pierre Macherey's (1998) phrasing, philosophy as an operation is practice itself, "as it puts back into question the limits inside which its activities are carried out." Theoretical curiosity as philosophical practice and operation is a continual operation that unsettles the quotidian, and the grounds of ideology as embedded within existing templates of objectivity and subjectivity. For Macherey, such a philosophical practice "is the movement of going past limits, of reflecting the immediate in the mediated" (pp. 36–37). It is a practice that intervenes within the sometimes volatile, sometimes lack of determinate conditions in inherently antagonistic social relations of capitalist hegemony. It is a desire to realize the opportunities presented within unanticipated occasions and historical conjectures.

Theoretical curiosity always takes form in material practices that arise from actual and contingent social relations. As a collective practice informing individual subjects and the projects they participate in, it eludes the lures of an irrationalism that would champion absolute indeterminacy, chance, and the calculus of the possible. It also resists well idealist and monologizing nets that recuperate it, drawing it down for normative liberal or social democratic critique, partners in the interests of instrumental reason. Foucault delineated ways by which "dispersed knowledges" are normalized, including disqualification of what is officially sanctioned as useless knowledges. This is the normalization process itself by which forms of knowledge, their proponents, and their specific mediations and transmission become interchangeable, reproducible, and generalized. This is evident in the infamous standardization of the French pedagogical system and in current American "progressive" educational policies implementing the rationality of accountants into the daily experience of students and teachers. In all instances normalization performs the same work as the "general equivalent" of monetarization, as performance based, quantitative measurement guides the corporate hierarchies and their "pyramidal" centralization of these "subordinated" knowledges for effective management. The varied disciplinary and regulatory mechanisms that normalize such subordinated, dispersed forms of knowledge, for Foucault, compose the structural armature of the modernist *episteme* in which theoretical curiosity operates. In Foucault (2003), these mechanisms correspond to the permeability of military and political regimes and their maintenance of order whose legitimacy, deploying knowledge, is often reducible to "who has the weapons" (p. 158).

As a punctuating point for critical reflection, the everyday situates a problematization of our time and its episteme. For Kristin Ross (1987), "everyday life harbors the texture of social change. To perceive it at all is to recognize the necessity of its conscious transformation" (p. 4). The critiques of modern urban daily life discussed here, as formulated by Lefebvre, Debord, and others, present ways of distancing from the grain of "facticity," of the givenness and resistances of circumstance, necessity, and repetition. My purpose in this chapter is not to render a historicist account. I do profile a distinct conjuncture whose leading agencies and determinate conditions can never be fully discerned. The discussion of the intentional interceptions of theoretical curiosity of Althusser, Debord, and Lefebvre is intended as an analogue of the problematics and impasses of the contemporary North American left and a rethinking of the conditions by which the political is produced. This demands frank acknowledgment of the self-reproducing limits of existing interventions of the progressive movements of the left in America and cosmopolitan cultural milieu, whether localized in organizations working for environmental justice, school reform, health care, workplace contracts, or struggles for identity and recognition. This requires ending all pretence to notions of reflecting and representing their various constituencies and interests.

Theoretical curiosity is not intended to define a particular practice or enduring trait; neither does it delineate any given form of situated intellectual activity, training, or position. It neither mirrors everyday life nor exists autonomously, independently speculative as a mode of theoreticism living in the intentionally atemporal realm of Hegel's "spiritual animals." As a practice open to engaging and actualizing all resources, it works a temporality not confined to an abstracted reading of the nature and experiences of the "present." It raises the stakes for any vital future for the left and progressive struggles for social justice by scratching through surfaces and the often-unthought premises guiding all manner of current strategies and tactics. It attempts to address, what, in another context Negri and Hardt (2004) identify, in their problematic project of the "multitudes," as a "conceptual lack concerning what the Left is and what it can become" (p. 219). The practice of a critically sustained theoretical curiosity advocated for here would raise the stakes in confrontation with the absolute exhaustion of social democratic reform strategies, organizational structure, and agenda, including many "grassroots" movements of all stripes in their naïve claims of a "realist" activism free from or independent of ideological labor and theoretical practice. It is a calling to account for an intense and constant questioning of their grounding motivations and purposes for existing as movements.

My intent here is not to claim any strong mutual influences among these select authors whose differences with regard to key questions and commitments range the spectrum of the humanist and anti-humanist currents of postwar thought and practice. I do wish to offer a brief portrait of some of their shared theoretical field and problematizations in which experiments and innovations in Marxist practice became necessary and desirable. Implicit in postwar critiques is an encounter with the everyday as a vanishing point and limit to thought. Since Sartre, the unraveling of grand historical and speculative narratives, whether Hegelian, Marxist, or in the legacies of the Enlightenment reason, revealed the negative power of the myth of a progressive and totalizing rationality. Their critiques of the founding principles of modernist secular states share an intense recognition and rejection of the hollowness at the heart of systems of thought, including dominant and official canonical renderings of historical materialism and dialectics that frequently mirror capitalism's own abstracting rationality. For Althusser and Lefebvre the chance to intervene was in large part made necessary and possible within the PCF's "outstretched hand" policy of gradual and controlled de-Stalinization, in what increasingly came to be seen as an instance of class struggle within theory itself. As in the case of Merleau-Ponty, these "adventurers in the dialectic" and laborers of the negative, while in no way representative of autonomous theoretical productivity of the postwar French *gauchiste* movements, nonetheless are of ample interest in signaling some of its tensions and critical affinities.

The conceptual apparatus of contemporary thought remains ambivalently locked in an uneasy relation to the power and passion of theoretical curiosity. As a formative and initially unstructured aspect of attention, perception, and affect, it has remained marginal to the purposefulness and historical self-conscious collective mobilizations and projects into late modernity. Idealized as a timeless and disinterested virtue, it is more commonly trivialized, disparaged, and appropriated. For Martin Heidegger (1962), discussed further in Chapter 7, curiosity represented a "tendency" within the vicissitudes of being. Sharing a disdain for the everyday with contemporaries, Lukacs and Adorno, he claims that "curiosity is everywhere and nowhere" (p. 217).

As a characteristic of Dasein it is actualized in human experience as a continual engagement with "idle talk," the chatter of the mundane and the ontic. Rendering a modern, existential analytic reading to its classical, Platonist ontology, it is an instance of the modality of appearance and simulacra. As a diversion, covering up, or forgetting, it is also a "veiling," flight, or veering away from Being.

Theoretical curiosity interleaves exteriors and interiors, surfaces and depths, remembrance and futurity. It tracks thought's involutions—the spacing and pulse of thinking, the emergence and dissolution of percepts, notions and concepts. As a tendency, *pace* Heidegger, it is thought's own temporality, placing and displacing experience and knowledge in the quotidian world. It is the restlessness of the experience of time and negativity, a conduit for continuities and dislocations alike. In challenging Heidegger's dismissal of it as a flight from "authentic" experience, projects like those initiated by Freirean critical pedagogy begin within the textures of daily experience under present social conditions. A care for being impels a reordering of the nature of lived experience. As the object of disparate revolutionary projects, such as those of the Popular Front, the Surrealists, Situationists, and critical pedagogies, an untaming of this otherwise-innocuous passion could constitute a shifting of the known coordinates of the phenomenal world, including the production of social space and time. The passions untamed could become the eloquence of everyday life transformed.

The Everyday

> By lorries along Sir John Rogerson's quay Mr. Bloom walked soberly past Windmill lane, Leask's the linseed crusher's, the postal telegraph office. Could have given that address too. And past the sailor's home. He turned from the morning noises of the quayside and walked through Lime Street. By Brady's cottages a boy for the skins lolled, his bucket of offal linked, smoking a chewed fagbutt (Joyce, p. 71).

The notion of "irreducibles," traces, or resistant remainders is important for a critical rethinking of social space. The irreducible or trace presents the resisting power of objects that do not correspond with subjects in relations of identity. The language of contradictions within a sedimented history of the dialectic is replenished. Traces refer to those discrepant oppositions and anomalies that, crystallized in the quotidian experience of whole social formations, jab at the smooth surfaces of a social totality and its triumphal claims to progress. In Henri Lefebvre's (1971) formulations, temporality and spatiality are contingent, capable of being fundamentally realigned. The administered life is never absolute, and agency reappears in the unexpected, in the irreducibles. He claims that

> if the circuit is not completely closed it is not for want of purposes or strategic intent but only because "something" irreducible intervenes, "something" that is perhaps Desire, or Reason(dialectics) or even the City... The only way to stop the circuit from closing is to conquer the quotidian" (p. 73).

"The everyday escapes," Maurice Blanchot (1993, p. 244) suggests. In response to Lefebvre, Blanchot cautions that phenomenal experience resists any detective work and all processes of unconcealing. The everyday eludes system, representation, and conceptual rigor. It is the expanse and horizon of non-reflective experience. In its arbitrary but usually unquestioned regularity, it is the domain of the common-sense order and rhythm of a world with others. For thinkers like Barthes, Blanchot, Lefebvre, and Benjamin it takes shape in the blur of the familiar, characterized best perhaps by street traffic, conversation, noise, graffiti, handshakes, lights and signs, all often anonymously choreographed in their patterns. An undifferentiated subjectivity, knowledge, and objectivity occur in the absence of critical awareness of the historical origins and structures of daily life.

In Bradley Macdonald's (1999) reading of the place of the "figure of desire" in Marx's *Economic and Philosophic Manuscripts of 1844*, "desire represents the constant striving of historically embedded humans to overcome their socially constructed limitations and strive toward their human potentialities" (p. 28). Lefebvre, who first translated this early treatise of Marx into French, was profoundly influenced by its ontological orientation. For Marx (1992), the human condition is characterized historically by distinctions between thinking and being, between what has distinguished modern variants of idealist and materialist thought. As a natural being, "as a living natural being he is on the one hand equipped with natural powers, with vital powers, he is an active natural being; these powers exist in him as dispositions and capacities, as drives" (p. 389) and "passion is man's essential power vigorously striving to attain its object" (p. 390).

Desire and curiosity intersect in the contingencies of collective life. As intensities of experience, they are congruent but differentiating as degrees of care and a detecting of difference. Theoretical curiosity, working the edges and absences in everyday life, attends to desire's savor of the sensuous, the grain and textures of things, notions and images that arouse its own continuity. Desire and curiosity conjoin in daily life, in traces of their utopian moments. It fosters recalling imaginary worlds hospitable to pleasure, to an economy of difference that is not cumulative and accountable in situations and moments of rupture of the quotidian where exuberance and creation arise. Desire as a generative principle of life is perhaps the ancestor that travels with curiosity in its errant flights and investigations; together, as irreducibles, they are fonts of abundance contained within historical experience and memory. They gather the kindling for the building of future fires, in the taking back of time and urban spaces to compose another world here.

Evident in the surfaces and textures of urban life, cultural and political modernity offers itself to theoretical curiosity a critical reflection on historical being and the composition of everyday life. Jostled in the scramble of urban transport, or observing street life in the poses of a hawker, a hustler or flaneur, a critique of everyday life can only emerge from the claim of things on us. A critique of modernity and modernism questions the given relations of everyday life, including their transparent order, design, and patterns, in all their symmetries, dissonances, and arbitrariness. The everyday, unreflected in the "naïve" or natural attitude, remains an undercurrent within the elusive nature of collective experiences of modernity. In Lefebvre's critical approach it summons a simultaneously intensive and extensive reading of the phenomenological screens of the spectacle of the ordinary in its continual repetition and difference. For Lefebvre, the power of the everyday requires a critical labor of the negative, the qualitative change of daily life offering up connections formative of subjectivity and historical agency. It is the actuality of historical desire in its "ability to create in terms of everyday life from its solids and its spaces—to make something lasting for the individual, the community, the class; the reproduction of essential relations" (p. 35).

Lefebvre sustained an ongoing examination of the everyday and the disparate significations of the collective experience identified as cultural modernity. Arguing that the quotidian and everyday were invented with the rise of industrial capitalism, he proposed a reinvention of a "style" counterposed to the reproduction and repetition of an everyday still dominated by antiquated, repressive cultural practices. A style designates ritual aspects of premodern or precapitalist organic communities, social relations and integration into the workings of cyclical time, and can include styles of power, cruelty, and wisdom, in techne organization and the productive skills of a world of experience without the quotidian. The alienation of workers from control of their productivity, wage labor, and the compulsions of consumerism have all contributed to the creation of a world restricted to quotidian relations wherein objects, appearances, and possessive individualism reign sovereign. In seeking release from the depths of alienation, social mobilizations can realign temporal and spatial relations. They reset the clocks and calendars of modernity to release the productivity of thought that would actualize the "extraordinary in its very ordinariness" within the everyday. As a strategy for forays into "experimental utopias" he described that

> the method employed is then that of imaginary variations on themes and exigencies defined by the real as understood in the broadest sense: by the problems posed by reality and the virtualities held within it... it navigates between pure practicality and pure theorization... We could give the name "experimental utopia" to the exploration of human possibilities, with the help of the image and imagination, accompanied by a ceaseless criticism and a ceaseless reference to the given problematic of the "real" (Lefebvre, in McDonough, 2009, p. 105).

For Lefebvre, writing without the ability to foresee the Events of May 1968, this meant recreating the communally based "festival and gather together culture's scattered fragments for a transfiguration of everyday life." As a tenet of a left militancy, unrestricted by dogmatic and reductive economic determinism, the intent of projects inspired by a "cultural turn" constitutes a revolution of everyday life. For Lefebvre, it bears the banner for a new social order freed of the systematic reduction in the quality of everyday experience. Proclaiming the arrival of the Festival, he writes:

> The revolution of the future will put an end to the quotidian, it will usher in prodigality and lavishness and break our fetters, violently or peaceably as the case may be. This revolution will not restricted to the economy, politics and ideology; its specific objective will be to annihilate everyday life: and the period of transition will also take on a new meaning, oppose everyday life and reorganize it until it as good as new, its spurious rationality and authority unmasked and the antithesis between the quotidian and the Festival—whether of labor or of leisure—will no longer be a basis of society (p. 37).

Lefebvre was a self-styled "utopian" Marxist. In rejecting a totally administered life, a "bureaucratic society of controlled consumption," he promoted the creation of "moments," situational interventions into the tangled sameness of ordinary life in cosmopolitan cultures. Specifically, his philosophical and political projects included interventions into the "closed circuit" comprised of production-consumption-production, in which "desires are run to earth." His stated intent was to "prove the existence of irreducibles, contradictions and objections that intervene and hinder the closing of the circuit, that split the structure" of production and consumption in the formulation of a new unity, or "style" of life reminiscent of the organic totality of communal life in medieval and precapitalist social patterns (p. 75).

Lefebvre's reading of everyday life as the place of transformation encourages a materialist understanding of desire, or what in Marx constitutes humanity's "vital powers," "dispositions," and "passion." The intent of his critique of the quotidian is in large measure a historical re-reading of Marx's theory of accumulation and alienation. It is also a stance against the sociologists and political economists of postwar Europe, who would, often inadvertently, represent a return to Hegel's invocation of the State as *geist*, the spirit of history incarnate. "Will this age witness the triumph of Hegelianism and of the totalitarian state rather than achieve the philosophy of a human totality?" (p. 56) he asks, remarking on the incomplete rationality and instrumentality of techniques for administering human desire, productive power, and labor, in both capitalist and state-socialist countries.

For Marx, objectification and alienation are not continuous or inescapable aspects of a fundamental ontological condition, as influential interpreters of Hegel, like Jean Hyppolite (whose seminar Althusser attended), claim. They are better thought of as the specific reified character of one form of objectification under the reign of a world dominated by economy. In Macdonald's (1999) formulation, a

> materialist conception of desire potentially engenders a way of understanding all the ways in which desire and pleasure become invested in everyday life, be they practices related to the capitalist economy, patriarchy, compulsory heterosexuality, or other spheres of power that reside within our life-world and are neither clearly nor necessarily reducible to the economy (p. 32).

In the pre-war context, the fleeting luxuriance that Walter Benjamin's (1999) personae, including the Idler, the Student, and the Gambler, partake in is one that beckons and flickers at urban edges, the *banlieus* of the capitalist spectacle. They carry on their non-productive activity drifting at the peripheries of a Calvinist economy of austerity within abundance, of consumerism and want, libertinism and moral order. Prefiguring the Situationists own *derives* and "psychogeographies" of the city, the Idler shares the singular discovery and pleasurable exploration of multiple and intersecting urban times with the Student and the Gambler. Existing on its own peculiar mappings of urban terrain outside the circumscribed instrumentality of productive, alienated labor and utility, the Idler is personified by the flaneur in an "unlimited duration which fundamentally distinguishes it from simple sensations of pleasure" (p. 806).

Unlike Benjamin, however, for whom the irreducible, as an excess that has not been absorbed into the totality of capital reifications of labor and commodities, and for whom it requires a conceptual labor of decipherment, Lefebvre's cross point, chiasmus, is the place and non-place where philosophy encounters its other. "Everyday life is the object of philosophy precisely because it is non-philosophical" (p. 17). This encounter, a tension in becoming, is also a frontier for theoretical curiosity. It brings up to the surface the resistances brought to bear upon any conceptualization of the cohesive structuring of society organized through the mediation of objects, as commodities, whose value and valences are reduced to their traffic in exchange.

Lefebvre renders Marx' s insights, particularly the influence of the *Economic and Philosophic Manuscripts of 1844*, into contemporary cultural politics. In the hypermodernized France of the 1950s and 1960s, which witnessed the aggressive transformation of its society to an American-style corporatization and consumerism, and, as Kristin Ross (1996) documents, one exemplified by cars, movies, and household cleaning goods, the everyday life, familiar for generations, never returned with the ending of the Occupation. Lefebvre brings into focus the repetitive nature of the everyday, the quotidian as a rhythm appearing outside of time, in the cycles of alienated urban lives in general, but in particular the lot of women in traditional, patriarchal family economies. For Lefebvre, the misery of the quotidian is measured in the weight borne by women in

> basic preoccupations with bare necessities, money, tradesmen, provisions, the realm of numbers, a sort of intimate knowledge of things outside the sphere of material reality: health, desire, spontaneity, vitality; recurrence, the survival of poverty and the endlessness of want, a climate of economy, abstinence, hardship, repressed desires, meanness and avarice (p. 35).

In the new modernization that affected France more quickly and intensely than other European countries after the war, a new privatization reinforced perennial gender partitions within a previously unimaginable economy of consumerism. Daily life is newly configured in intensified divisions between public and private lives, between sexuality and technique, politics and domestic concerns that also witness a disciplining of these divides. Shopping, fashion, and the care of the house constituted the daily lives of women, and in the new magazines, such as *Elle* and *Esprit*, the domains of romance, fashion, beauty, cooking, and practical advice comprised the circumscribed content. As Kristin Ross notes, "career, politics, science and economic information are relegated, in a strict gender division of access, off limits" (p. 85).

Perhaps Ross's call for transformation of everyday life is an invocation of a new image of Pandora. The legend of the first woman, fabricated by the gods in reprisal for Prometheus's theft of the first technology of fire, suggests a critical agency within the confines of daily life. Laura Mulvey (1996) reads the myth closely, tracing how the curiosity associated with her looking into her vase delineates a long historical association of women's relation to enclosures, to secret or forbidden places. Laura Mulvey regards Pandora's gaze an "active look," one that discerns relations between interiors and exteriors. Her fateful gesture, in opening up the container, "may be read as curiosity about the enigma of femininity itself" (p. 61).

The skeins of everyday experience are material in their patterns, habits, and relations. Lefebvre's ongoing critique of everyday life traces out to some degree the fortunes of dialectical materialism in the postwar period. As Diana Coole (2000) observes in her commentary on Adorno and Merleau-Ponty, it is a "juncture where negativity was becoming associated with a wilder process of an anti-rationalism which distanced it from its Hegelian and Marxian formulations, while its importance for an emancipatory political project and rationality was yet . . . retained" (p. 122). As desire is actualized in an everyday that it flees, so too curiosity inhabits the quotidian, usurping conventional discourse, structures, and habits of mind on its uncharted, capricious, and nomadic routes. As the work of the concept, in dialectical language, it is for Hegel and his materialist interpreters, an active absence, the virtuality, or plenitude immanent to historical existence in its unfolding.

I wish to claim a kinship for theoretical curiosity and the power of the negative. For Hegel "thinking is, indeed, essentially the negation of that which is immediately before us (Marcuse, 1960, p. vii)" or, as Paul Valery phrased it, "thought is the labor which brings to life in us that which does not exist" (in Marcuse, p. xi). It is a historical desire, given shape by the contours and conventions of daily experience. Theoretical curiosity, as a political talisman, propels thinking and practice as operations within a field of powers and resistances, of always questioning and engaging with and against a facticity that would eviscerate and reify historical consciousness. For Coole, "negativity and critique are intimately related, indeed almost synonymous, since negative thinking is immanent critique" (p. 55).

As a social force critical curiosity is thought of as a form of material power of relays between bodies and minds. It is what Coole describes as a "vibrancy" and "multiplication of connections and relations, oppositions and difference" formative of new social formations (p. 59). Its acts are tenuous, uncertain, working a negativity that would steal into the solidified domains of disciplined and absolute temporality and spatiality. It is the material, rhizomatic power of thought that can actualize the virtuality of a political imaginary. It works toward what Deleuze (1994) asks of thought's responsibility when he questions "to what are we dedicated if not to those problems which demand the very transformation of our body and our language?" (p. 192).

For Lefebvre, philosophy's limit and pedagogical purpose is "practicing a sort of maieutic in assisting the birth of everyday life's potential plenitude" (p. 18). In Spinozist fashion, it is an affective-epistemological power between bodies. As a critical project this power is a singularity and intensity within everydayness, the potential plenitude or virtuality that could sustain a qualitative and collective leap within the everyday in its transformation. This excess resists and contests the constitutive forces of reterritorialization. Projects and practices sustaining and sustained by a theoretical curiosity deploy the productivity agency and material resources held in reserve and compromised conditions within the everyday. These are the grounds for a collective, phenomenal becoming that is always "ontologically given through the autonomous self-positing of human beings" (Roberts, 1999, p. 23).

Striking the Matches

> Cityful passing away, other cityful coming, passing away too: other coming one, passing on. Houses, lines of houses, streets, miles of pavements, piledup bricks, stones. Changing hands. This owner, that. Landlord never dies they say. Other steps into his shoes when he gets notice to quit. They buy the place up with gold and still they have all the gold. Swindle in it somewhere. Piled up in cities, worn away age after age. Pyramids in sand. Built on bread and onions. Slaves. Chinese wall. Babylon. Big stones left. Round towers. Rest rubble, sprawling suburbs, jerrybuilt, Kerwan's mushroom houses, built of breeze. Shelter for the night (Joyce, p. 164).

Seeking the openings and moments when change seems imminent, an insurrectionary spirit takes advantage of their rarity to act. In taking Rimbaud's cry to "change life," avant-garde interventions of the twentieth century, particularly in their dadaist and surrealist incarnations, were intent on the transformation of daily life, one that would merge the aesthetic with the quotidian, the marvelous and the mundane. These movements were resources for the Situationists International's (SI) formation in 1957, the groupuscule composed largely of former or continuing artists and intellectuals who vowed to bring the surrealist project to fruition in a supersession of art itself as it melded with everyday life. The Situationists were key to the energies released in the Events of May 1968, through their employment of dispersive strategies and in tactics promoting a "propaganda of desire." They proposed an end to all specialization of knowledge and activity, including art in a world in which estrangement would be replaced by a "thoroughgoing fusion of reason and passion." The short-lived strategic alliance between Lefebvre and SI's most audible provocateur, Guy Debord, is evident in their program: "we still have to place everyday life at the center of everything...everyday life is the measure of all things: of the fulfillment or rather the nonfulfillment of human relations; of the use of lived time; of artistic experimentation; of revolutionary politics" (in Plant, 1992, p. 63).

Reinvigorating the pose of wandering flaneurs, the Situationists advocated a new urbanism, sharing concerns also articulated throughout the later writings of Lefebvre, Debord, Vaneigem, and the SI, in their publications as well as their a creative actions. They devised strategies for new ways of knowing and experiencing city life. Debord (1994), author of *Society of the Spectacle,* articulated a vision of unestranged experience. Influenced by the Surrealists as well as by Georg Lukacs, the musings on a "spectacular" society, written in 1967, offer a prescient analysis of the commodification and containment of all human experiences, including dissent and radical gestures, through the administration of capital's techniques of managing desire. Amounting to a manifesto, he declared

> Of arms and the man the spectacle does not sing, but rather of passions and the commodity...the commodity's becoming worldly coincides with the world's being transformed into commodities. The commodity's mechanical accumulation unleashes a limitless artificiality in face of which all living desire is disarmed, comprising a seamless falsification of life (p. 44).

The *derivé,* the technique of "locomotion without a goal" was intended to reawaken elements of mystery, eroticism, and surprise into the experience of perambulating the city's streets. As Sadie Plant explains, the derives were one aspect of an effort to promote a "psychogeography," a revitalized urban life in situations, events, and moments conjoining architecture and passion, the contours of streets, lighting, and voices with the affective intensity of individuals. Inspired at first by the CoBrA group of artists, and, in particular, by the work of the Dutch utopian architect Constant Nieuwenhuys and his designs for a New Babylon, the *derives* would make one notice "the way in which certain areas, streets, or buildings resonate with states of mind, inclinations, and desires, and to seek out reasons for movement other than those for which an environment was designed" (Plant, p. 59).

In their experimental militancy against lives lived as spectators in contemplative passivity, Debord and the Situationists advocated ways to merge theory and practice. Individual and social, private and public aspects of estranged commodified existence would be combined, insisting that "revolutionary organization must learn that it can no longer combat alienation by means of alienated forms of struggle" (Debord, 1994, p. 126). The new urbanism of the *derivés*, of accelerating and suspending situations, spaces and moments, would constitute an active, communal "critique of human geography." It would be located in multiple, shifting places of play, by "virtue of freely chosen variations in the rules of the game." Ludic spaces and temporality, freed, even if fleetingly, from instrumentality and divisions into parcels of labor and leisure, of productivity and stupefaction. It was a concerted, anarchic attempt to "construct places and events commensurate with the appropriation, no longer just of their labor, but of their total history."[1] In a Benjaminian perspective, it is the world of the Arcades, the luminous objects of desire and distraction, existing in their own spectral and luminous glow.

Grounding a theoretical curiosity, the Situationist's strategic repertoire also proceeded to resituate existing dogmas and practices through the process of *detournement*, calling it "not a negation of style, but the style of negation" (Debord, p. 144). Debord and the Situationists violated intellectual propriety by intentional plagiarisms and appropriations that reversed or inversed meanings. At the linguistic level, Debord speaks through the "reversed genitive," citing Marx's own turning around of Feuerbach's "philosophy of poverty" into the "poverty of philosophy." In claiming this rhetorical tactic as the "antithesis of quotation," Debord intended a critical, reflective distance to "whatever has been turned into an official verity." It sought a rupture with totalizing discourses, and the fixing of theory in monologic, ahistorical perspectives that deny their interests and positionality. *Detournement* is a form of intellectual force that "mobilizes an action capable of disturbing or overthrowing any existing order" through parodic inversion, destabilizing binary divisions governing thought and experience. Drawing inspiration from Lautrémont and Rimbaud, the Situationist's linguistic and voluntaristic performances, including the writing of an alternative dictionary, were intended to fulfill the revolutionary promises of avant-garde experimentation through practices of incitement and provocation. Most importantly for Debord and the Situationist's was their determination to end the contemplative existence of spectators of the spectacle. The intellectual and revolutionary vanguard would reanimate Marxian intent and praxis as a "reminder that the existence of the theoretical domain is nothing in itself, that it can only come to self-knowledge in conjunction with historical action" (pp. 145–146). Detournements were claimed as their "signature," emblematic of an interventionist spirit seeking, and, very briefly finding, in Europe's largest twentieth-century general strike of ten million students and workers, a carnivalization of daily life.

Instigative critiques of everyday life such as those advanced by Lefebvre and the Situationists adamantly opposed institutional and reductive theoretical dogmas and programs, whether doctrines of state bureaucracy and the new technical intellectuals of postwar Europe or those, like the PCF, that they viewed as hobbling the potential for revolutionary action. The European Marxism inaugurated by critical projects—such as Lefebvre's fifty odd years of concern with developing a sustained critique of daily life under the conditions of capitalist modernity—labor under the solar glow of an idealized totality. In Lefebvre, as well as in Debord, this ideality is that of a robust and authentic humanist subject. Disregarding the unreflected masculinism of their vision of the "total Man," despite Lefebvre's attention to the intensity of privation and privatization of oppression in the everyday life for women, they often assumed, uncritically, the coming to fruition of a fully present, autonomous subject released and liberated from the trammels of capitalist alienation.

The divisions in the spatial and temporal aspects of life could be reintegrated into a seamless unity of sensuousness and intellect. Their work was motivated by an abiding critical curiosity for ways of trespassing, not only the estranged conditions of labor under capitalist relations of production, but of entrenched and ultimately collusive work of intellectuals and philosophers and theoreticians of all existing alliances. Their theoretical ambitions, flaring most brilliantly and unexpectedly in the Events of May, would seek to bridge the seemingly intractable chasm between thought and action, intellectual and manual labor, speculative passivity and material, active intervention, reflection and practice.

Experiments in the Aleatory

> Ineluctable modality of the visible: at least that if no more, thought through my eyes. Signatures of all things I am here to read, seaspawn and seawrack, the nearing tide, that rusty boot. Snotgreen, bluesilver, rust: coloured signs. Limits of the diaphane. But he adds: in bodies (Joyce, p. 37).

To recognize of the aleatory, the indeterminate and radically contingent, is to acknowledge and work with the simultaneous arrival of multiple converging forces, alliances, and powers. The aleatory transmits currents of an unpredictable theoretical curiosity, providing the times and places for agency and intervention in the course of history. The element of chance is a power in and for social change. It eludes any productivist teleology of progress, whether in capitalism's technological development or in the revolutionary praxis of Hegelian Marxism. It is a subordinate, if not heretical, discourse within any materialism. For Lefebvre (1995) this is formulated clearly in his writing in the aftermath of Khruschev's revelations of the traductions of socialism committed under Stalinism. In his perspective

> the new, the unforeseen, comes into being from inside that old bag of tricks known variously as destiny, man's history or prehistory, chance, historical determinism, economic determinism. Just like a ship history can drift off course (p. 37).

A critical curiosity regarding the material forces latent in the contingent attends to the folding and unfolding of historical conditions, their pliability, or susceptibility to provocation. In the texts and comic book detournements, in manifestos, pamphlets, and proliferating graffiti, as well as in their *Internationale Situationiste,* provocateurs like Guy Debord, Raoul Vaneigem, Michele Bernstein, and Mustapha Khayati sought very specific occasions to catalyze awareness and resistance to the society of the spectacle in all its alienating tawdriness, elegance, and ubiquity.[2]

The Society of the Spectacle constitutes a false totality. A defaced surface to all things and reified relations, it constitutes nothing less than a life without depth. Outside of embodiment, it is an image of image, as they flicker across twenty-first century screens; digital technics think through us in cyber-mediated relations. The spectacle is a society glinting with backlit screens without actual faces in a continually flickering "interface." The Situationists tried to pry behind the surface of the everyday to reveal the *tain*, the silver paint behind the mirrors of the spectacle. Their theoretical practice was a ludic investigation for a realm of freedom below the historical and oneiric layers of endlessly repeating sound—images, and beyond the ghosted grey of capital. For these small groups, situations were intended to facilitate a revolt against conditions of alienation in every aspect of daily life, claiming "we will only organize the detonation" (Knabb, 1981, p. 113). A revolutionary experience in turn would be one that constantly generates situations, festivals, new uses of space and time, occasions for transgressing the terrain of capitalism's reign of private property as well and its psychic taming of the passions within and between classes.

Within twentieth-century Marxism, an interest in the aleatory can be discerned as early as in Rosa Luxemburg and the Spartakists in Germany. As Michael Lowy (2000) observes, "with her famous slogan, socialism or barbarism, she broke, in the most radical possible way, with all deterministic teleologies, proclaiming the irreducible factor of contingency in the historical process" (p. 39). Lukacs regarded this interest, the *augenblick*, this "blink" or the "glance of the eye," a detecting of what, for Zizek (2000), is the singular historical

> moment when, briefly, there is an opening for an act to intervene in a situation—the art of seizing the right moment, of aggravating the conflict before the system can accommodate itself to our demand (p. 119).

In Zizek's remarks, this reflects a Gramscian interest in contingency and conjecture. It is also a point of convergence for the politics of the gaze. Theoretical practice, regarded as forms of seeing, from the classical era through to Heidegger, Lukacs, and Debord, remains a point of contest. Vision, as the primary channel and sense of representation sustains theoretical interventions in many political projects. The *augenblick*, discussed in the context of Heidegger and Merleau-Ponty in the following chapters, also invokes visualist tactics, which, however, are not distancing and speculative. It is a marker for an event, for the contingent, as I will illustrate in the case of the formation of the Events of May 1968. It represents the unpredictable, the virtual, and possible transformations in the history of perception and the structures of everyday life. Rejecting historical agency reduced by Party theoreticians reading their *Diamat* tea leaves of existing historical conditions, this art of the "glimpse" waited expectantly for signs for action, encouraging eruption, the flare of Events without evident precedent. Zizek (1999) appositely and ironically quotes Lenin when he stated, "history will never forgive us if we miss this opportunity" (p. 39). A sphinx blinks.

Within Heidegger's project of renewing metaphysics, curiosity or care would be a returning to ontology in placing our theoretical "desire to see" toward the future. As a volatile temporal site, he considered it as a point of convergence for a renewed alignment of philosophical desire. This "blink of the eye" is characterized for him by what McNeill (1999) finds as its two fundamental traits:

> the trait of finitude, whereby it marks the open, the sudden, unmediated, unforeseeable irruption of beings into presencing, and the trait of being held in such openness in such a manner as to be delivered over, always already to historical time, to being with others, and the to the claims of a tradition (p. xi).

An active theoretical curiosity for the aleatory element in the edifice of capitalist rationality reappears within the proposals for a "new materialism" by Althusser, Lefebvre, and the Situationists. As potential alternatives to the centralization of the Comintern led Parti Communiste Francais (PCF), articulations of a critical curiosity worked against the grain of a bureaucratized, economic reduction of historical materialism. Their open-ended theoretical and dialectical speculations on indeterminate historical processes were important aspects of a distinctly French cultural history and politics. French *gauchisme,* also evident in the projects of the *Socialisme ou Barbarie* group, with which Debord was briefly affiliated, was a renegade historical materialism. They refused official *Diamat* priorities and pieties. The cunning of their collective theoretical interests could never accept doctrinal determinism in any form. They all shared a radicalization of social theory that would surpass static analyses of the contradictions of alienated experience in capitalism to either economic forces alone or in mechanistic base/superstructure reductions with their attendant reflection and correspondence explanatory models of culture.

The writers and contributors to *Socialisme ou Barbarie*, including Cornelius Castoriadis, Claude Lefort, and Daniel Blanchard, who constituted a split within the Trotskyist movement, were influential at the time in advocating self-management on the model of the revolutionary councils of the Spartakists. Anselm Jappe (1999) points to their critical engagement with the traditional Left in their opposition to its myopic advocacy of objective productivism, concerns about working hours, leisure time, the provision of better educational institutions and social services. Their advocacy of the "subjective" and "poetic" resonated with the stirrings of the Situationists and their intentionally fleeting alliances, in their dissemination of the possibility of self-management in all aspects of life. Jappe comments that these writings cumulatively argued that the true content of socialism was neither a planned economy, nor simply an increase in the material standard of living, but rather the prospect of giving meaning to life and work, releasing creativity and reconciling man with nature (p. 92).[3]

For Althusser, the aleatory is constituted in a philosophy of "encounter." As a distinct reading of materialism, it stands "completely opposed to the various registered materialisms, including the materialism commonly attributed to Marx, Engels and Lenin, which, like every materialism in the rationalist tradition is a materialism of necessity and teleology" (in Mulhern, 1994, p. 40). For Althusse, taking his cues from Machiavelli, theoretical practice is a crucial site for class struggle within Marxism. It is a continual war of positions, of insistent antagonism. A "thinking in the extreme," it is a pedagogy of theoretical curiosity's encounters, of taking chances in an imperiled world. It is an encounter where the appearance of the fox, rather than the lion, becomes the emblem of a rigorous engagement with the materiality of conceptual labor. This meant also a reconsideration of the elements of contingent relations for both linking the politics of the PCF to a mass base as well as the unity of theory of practice through concrete analysis.

It was the figure of the Prince, and an enigmatic Machiavelli in his *virtu*, who becomes here a theorist of "concrete conjunctures." For Gopal Balakrishnan (2000) this interpretation makes the Machiavelli of *The Prince* the crafter of an

art of thinking focused wholly on the conditions of undertaking tasks immediately at hand, without anchorage in any underlying movement of history: a supposedly deeper, albeit more unstable kind of knowledge (p. 159).

While, as Balakrishnan indicates, "neither a monistic ontology, nor a claim about the primacy of the economic in history," Althusser's language of the aleatory is one of a vigilant alchemist, in which "concealed vectors" are capable of suddenly igniting "immanent possibilities" in a volatile present. A capitalist, structural "totality" was overdetermined by its own internal contradictory forces. Overdetermination became a base for his thinking of the possible and the role of a random interception into its functioning. It is a thinking of unfinalizable project of human community, of a world without origin, formal causes, or determinate ends.

Theoretical practice is here a thinking of the singularity of theory's own production. Theoretical curiosity as practice is co-determined by its outside, the everyday and contingent as the necessary crossing point for parallel, autonomous series or attributes of a singular encounter of "heterogeneous orders." It is, as Jean-Claude Bourdin (2000) describes, a working through of a distinctive materialist practice committed "to thinking randomness, the uncertainty of the events of the world." It is a materialism freed from the idealist foundation of the principle of sufficient reason, of having to explain itself on the basis of the relation between thought and being. It sets into motion the

idea of process without subject or completion, the affirmation of the necessity of contingency and the contingency of necessity, the possibility of things being rooted in the concept of an original disorder, the constitution of the world as singular events (p. 284).

For Althusser it comprises revolutionary theory's own indeterminacy as a praxis, operating outside of historical conditions of revolutionary necessity, circling over the City with its one-winged dialectic.

Althusser depicts a relatively indeterminate historical subjectivity, one materially produced in the circulation of discourses. These discourses, social practices, and distributions of knowledge and power are, *pace* Spinoza, considered immanent to the constitution of a subject. Key to dialogues between Deleuze, Foucault, and Althusser was a resituating of the place of the subject as a function of ideology. For Foucault and Althusser, in distinction to Lefebvre and the humanist, mimetic representational legacy adopted by classical Marxism, the subject has been dethroned, its pride of place in the humanist pantheon as a substantive consciousness and intentional agent of action sundered. In Warren Montag's (1999) incisive assessment, ideology is interrogated as a process by which subjects are materially inscribed, through specific discourses, practices, and disciplines. These expressions of power are symbolically imprinted; as writings on the body, they represent the effects of discursive processes of subjectivation. An intriguing question that remains, as Balibar (1996) poses it, is whether there can be negativity without a subject (p. 119).

Both Althusser and Foucault (who refused to use the word *ideology*) developed their influential projects as critiques of humanist historical representations of subjectivity. In both thinkers interiority and exteriority are displaced. A subject is produced and always active within force fields that suffuse and ripple through both mind and body. Interiority is an ideological space, the imaginary, secondary effect, and doubling of the materiality of everyday practice. Individuation and the partitioning of role, personhood, identity, and its concomitant values of conscience, intention, and will are thrown off of their Enlightenment orbit. Critical agency remains possible despite the formidable character of the processes of subjectivation that Foucault and Althusser delineate. In Deleuze's materialist ontology, and in Foucault and Althusser's elaborations of a political ontology, it is also a philosophy of difference. Reflecting on these critical theoretical premises is important for rethinking agency. If critical praxis floats in the unpredictable material currents of power it must also negotiate, then it is confronted continually by questions of how to act in the moment in the interest of political and cultural change. Agency informs political and pedagogical projects and is arrayed in its deployments by the very material productions of power that must often be resisted. In the Spinozist terms favored by Foucault, Althusser, and Deleuze, it is the power to act upon the actions of others.

The aleatory was staged in decidedly utopic gestures in the hands of the Situationists. They resisted Gramsci's cunning encounters with Machiavelli in which the conjunctural and chance elements for radical action are coordinated within a general understanding of the long, organic, and historical conditions for revolutionary practice. The Situationists, along with many of the groupuscules of the *gauchiste* sixties, veered toward a voluntarism and politics of spontaneity without mass base, organizational structure or a critique of the limits of subjectivity and desire under the realm of capitalism. In the writings of Raoul Veneigem this subjectivism emphatically marks an advocacy of personal or group lifestyle choices. These were what ultimately constituted forms of radical imposture or youth cultural practices, ranging from countercultural movements, yippie activism, or the foment of Dutch Provos that Debord and the SI came to condemn as recuperable to the society of the Spectacle. The Situationist's singular inspiration fused aspects of the Lukacsian "blinking eye" with surrealism's challenge to bridge the chasm between lived experience and artistic freedom. Theirs was a theoretical curiosity seizing social space and time against its commodification.

Inaugural Events

In a striking meditation on ethics and evil, Alain Badiou (2001) distinguishes between situation and event. Curiosity calls into question the limits of the knowable within existing existential conditions of experience and relation. For Badiou a "situation is composed by the knowledges circulating within it," while an event, which marks both a break and continuity with the situation, "names the void inasmuch as it names the not-known of the situation" (p. 67). An event of thought, as illustrated by Galileo, Newton, or Niels Bohr, or in music by the compositional structures initiated by Haydn, in twentieth-century social traumas, such as the Nakba or the Holocaust, and perhaps now in the aftermath of the World Trade Center demolition, in politics and ethics, marks a singularity, a rupture, and continuing, immanent break with the everyday. An unexpected interruption, an event appears without precedent and precondition.

For Badiou, an event, whether in science, in politics, in research, or in personal and intimate relations, is a "hazardous, unpredictable supplement, which vanishes as soon as it appears" (p. 67). It marks the inauguration of new "truths," as multiple, convergent forms of knowledge, not as static, monolithic ahistorical absolutes. In this sense, situations can never account for the emergence of the radically new or contain how they may usurp dominant practices, discourses, and ways of thinking and being. The power of such breaks, such as during and *after* the Events of May, exceeds the epistemological crisis of a social order's management of knowledge. The power of the "truths" that force their way into experience are propelled by disruption, by "violating established and circulating knowledges" (p. 70). Badiou's comments on truths as events recalls the passage from Plato's *Republic* in which Socrates questions whether "strange band" of those who are "lovers of spectacles" can be named philosophers. True philosophers, he asserts, are "those who love the sight of truth" (or, for whom "the truth is the spectacle of which they are enamored" (Plato, 1997d, 1102, [475e]). The force of an event invents new relations and terms for thinking, commanding a "fidelity" to the discourse it generates and imposes. This fidelity also generates what Badiou considers a "sustained investigation of the situation, under the imperative of the event itself" (p. 67). More than forty years after he continues to affirm that we are still the "contemporaries of '68 from the point of view of politics, the definition of politics, and the organized future of politics" (2010, p. 62).

The non-repeatable and singular activities leading to the events of May in France evidenced a dynamic convergence of multiple sources, sites of power, theoretical practice, and resistance. In its resistances and circuitous troubling of notions of progress, history, or economic determinism, a critical, negative, and theoretical curiosity might emerge as one of the vital forces and driving energies of social movements in their vision of an alternate present. France's imperial "maintenance of order" in Vietnam and, later against the NLF in Algeria, so vivid in the thought and practice of the generation of 1968, left no space for intellectuals and workers of that generation to remain spectators. Idle, speculative theory, as removed from partisanship in the issues confronting them internationally and nationally, galvanized millions to take sides, to become engaged, seeking alliances for the transmutation of the culture as a whole.

When historical conditions are favorable, theoretical curiosity can perform Hegel's "labor of the negative" as a catalytic agent, seizing the chances and rare opportunities that present themselves, in taking time back from the spectacular speeds of its consumption. That this catalytic action can prize power and the imagination was most remarkably demonstrated by the Events of May. Rene Vienet (1992), one of the members of *Enrages* at Nanterre in the winter and spring preceding, writes that

> the agitation launched at Nanterre, by four or five revolutionaries, who would later constitute the Enrages, was to lead in less than five months to the near liquidation of the State... Never has a campaign undertaken by so few individuals resulted in so short a time in such circumstances (p. 19).

The energies released by the events of May were also deployed within the proliferation of *Comites Vietnam de base (CVB)* groupuscules, organized by Maoist tendencies in opposition to their Trotskyist CVN counterparts. As Kristin Ross (2002) reports, these student splinter groups were not only active at the university level but most importantly within high school and community organizations. Among them was the UJC(m-l) that Badiou formed, one which Ross describes as a "Maoist group originating in December 1966 among the Althusserians of the Rue d'Ulm" (p. 89). A resolution from 1967 declared the necessity of forming a united front of youth against the intransigence of American imperial rule and the formation of revolutionary intellectuals "who will join with the workers and working people, who will institute new forms of organization that will make possible the realization of such a task" (p. 89). Students supported and joined in factory takeovers, replete with red flags never seen before, that often formed a unique expression of solidarity within families and communities.

Theoretical curiosity emerges in a surplus of experiments in praxis; the postwar theorists surveyed here, pursued, in their own distinctive and eclectic formulations, currents of thought whose origins derive outside of the Marxist canon, while reworking some of its central tenets and commitments. For Althusser, this critical theoretical curiosity and concern is evident in the exploration of a non-teleological, aleatory materialism, his actualizing appropriations of Spinoza and Machiavelli, and the attention the latter afforded him to think singularity, and the contingent conjunctures of *virtu* and *fortuna*. For Lefebvre, it is demonstrated by his continual renovations of Marxist theory after his break with the PCF, in his critiques of everydayness, modernity, bureaucratic consumer society, and his pioneering ventures toward a spatial politics and urbanism. For Debord and the Situationists, it is manifest in their modes of incitement, provocation, and attempts at puncturing through the fiction of totality at work in maintaining and managing the society of the spectacle. These practitioners and practices shared a restless and intensive reconsideration of the singularity of specific historical conjunctures. Each opened up alternative materialist interventions into the cracking fabric of the everyday in a tireless will to formulate new tactics and strategies within specific, often unpredicted circumstances and tensions of late capitalist development. The Events of May present questions that remain unsettled today in their intentions and attempts to actualize the potentials presented by history's dealing up of unexpected cards.

It is the Sphinx's question. As Badiou (2009) revoices it

> it is not by chance that Oedipus answers the Sphinx at the crossroads However, if he knows how to answer the question, What is Oedipus?, he must leaves as a (dead) remainder the question of this question: Who is the sphinx, having made Oedipus—man—into his problem? (p. 201).

What is this "man" that he is a problem for himself? The Events of May remain singular. As an "event" in Badiou's terms, they leave open an unfinished collective project. For Jean-Luc Nancy (2010), the era in which they appeared to flare out of nowhere, were and are still the beginning of a "present as the presence and co-presence of possible" (p. 16). It remains a virtual event, signed by the incalculable and a risking of time. It still signifies the task of starting to peer through the spectacle of late capitalism. Responding to all those who condemn and disparage its legacy, including President Sarkozy's attributions of decadence and decline resulting from it, Nancy asserts that it marks a beginning of a departure from the "age of world pictures" (p. 6) the "first irruption of the demand for a reinvention" of life (p. 8). In what was certainly an unprecedented convergences of forces, at once material, physical, and theoretical, it never comprised as a student, youth, or generational rebellion. For Nancy "we were exiting not only the time of 'conceptions', 'visions', or 'images' of the world, but the general regime in which a vision understood as a theoretical paradigm implied the sketching out of certain horizons" (p. 9). It was not "just a game of intellectuals but a feeling, a disposition, indeed, a habitus or ethos that entered into public consciousness and ways of thinking" (p. 10).

Championing interferences whenever possible, a critical theoretical curiosity maneuvers the conceptual, intellectual, and material resources, tools and inventions to intervene, pursue, and make multiple Foucault's "comings and goings." It is a call to constant reinvention and presentation of embodied coexistence. Burrowing beneath the endless reflecting surfaces of the spectacles of everyday life, theoretical curiosity would crack at the bad infinity of alienation; in the finally irrepressible desire for freedom in thought, action, and experience, it will pursue living substance behind its mirrors and glass. The rhythms and eros of embodied life will not be sterilized into the objects of any political representation attempting to deflect it. The stakes to be raised here are indefinite, uncertain, and perilous. It is nothing less than the human project to resist and thrive beyond all that hobbles the actualization of a genuinely social futurity and plenitude. Like Joyce's watchful and sometimes blinking Sphinx, we must remain open to what lies contingently before us, exposing the real, gaping yet hidden tears in the tent of the Spectacle.

Notes

1. Debord, author of *Society of the Spectacle*, articulated a vision of an unestranged experience. Influenced by the surrealists as well as by George Lukacs, the formulations of a "spectacular" society written in 1967 offer a prescient analysis of the commodification and containment of all human experiences, including dissent and radical gestures, through the administration of capital's techniques of managing and substitution of desire. "Of arms and the man the spectacle does not sing, but rather of passions and the commodity. . .the commodity's becoming worldly coincides with the world's being transformed into commodities," he writes, declaring the "commodity's mechanical accumulation unleashes a limitless artificiality in face of which all living desire is disarmed," comprising a seamless "falsification of life." G. Debord, *The Society of the Spectacle*, Zone Books (1994, pp. 42, 44). Debord, ibid., p. 126.
2. Mustapha Khayati, writing in a preface to a Situationist dictionary, claims that a new vocabulary will comprise a complement to the freeing of the proletariat that will give it voice. He compares their project to that of the *encyclopedistes*, maintaining that theirs will express the "qualitative and of the still-absent potential victory, the repressed of modern history (the Proletariat) and the return of the repressed. We are proposing the true liberation of language, for we propose to put it into practice free of all fetters." Captive Words, in T. McDonough (Ed.), *Guy Debord and the Situationist International*, October, MIT Press (2002, p. 179).
3. Jappe, *Guy Debord*, University of California (1999, p. 92). Jappe comments on Debord's readiness to engage in provocations that rework the torn texture and substance of the everyday and the spectacle. He observes that "throughout his adventurous life and continual conspiracies he was motivated not by ambition but the desire to rejoice in the drama of situations, to dabble with sets of historical contingencies" (p. 113).

Chapter 5
Curiosity and the Question

In this chapter I am interested in attempts to bring into relief the objects of curiosity's drifting, often errant, but not disinterested desire for knowledge, one that acts within an alternating, groundless multiplicity of possible objects, features, textures, rhythms, and intensities. Theoretical curiosity presents a cloud chamber in which the traces and collisions of thought can be observed as they radiate centrifugally from disciplined epistemological centers of gravity. I wish to substantiate my claim that theoretical curiosity always emerges from the phenomenal, perceptual world to which it gives expression.

Rethinking the sources and pursuits of theoretical curiosity is a rethinking of the freedom of thought itself. The critical curiosity I am sponsoring is engaged in making new relations between emergent ideas, perception, concepts, and representations. Springing up from the domain of the senses, rather than essences, it is a catalytic element of the desire for knowledge. It acts to bring into confluence affective, pre-conceptual, and pre-representational processes for explorations in thought, language, and discourse. As in William James's "perchings" and flights, curiosity is not thought's root but a rhizome. It arises within the configured mediations of an *episteme*, the historical organization and closure of knowledge. As a spur it can generate all manner of disciplinary research and their collaboration. Set into motion it works in their emergence within very material resistances; it presides as a kindly and tutelary muse within research initiatives, institutional structures, and relations of knowledge. It is a vital spring for the expression of potentiality within the incalculable productivity of any social ontology.

Thinking emerges to participate and comprise its own elemental temporal fields and configurations. Fields of thought, mobile and relational, are in continual encounter with existing canonical representations, structures, and relations of power. As potential points of resistance and support to disciplinary power, it vies with any orthodoxy and its sedimented institutional discourses. As counterweights to existing paradigmatic models and conceptual inertia, the adventures of theoretical curiosity pose a relentless interrogation.

In thinking through how certain inquiries begin and traverse shared experience, I want to emphasize how they are saturated by phenomenological qualities. Taking form in its alliance with the formative forces of imagination curiosity partakes a

M. Zuss, *The Practice of Theoretical Curiosity*, Explorations of Educational Purpose 20, DOI 10.1007/978-94-007-2117-3_5, © Springer Science+Business Media B.V. 2012

twilight or crepuscular experience, frequently marked by patterns, shadows, light, tactile impressions, sounds, and rhythms discerned and lost. The reflective trail that links any certain object of knowledge wanders; it is an inductive phenomenal path that can readily vanish. There is often only an inchoate sense that some experiences or research findings suggest a regularity, pattern, or symmetry (or asymmetry) that encourages the continuation of the search.

Ongoing and well-financed projects like genomics research, systems biology, astrobiology, and neuropsychology, for instance, discussed in Chapters 6 and 7, insinuate themselves in, and often against, the grain of conceptual order, category, and disciplinary boundary. This was illustrated in the last chapter, where I profile the convergence of disparate theoretical practices in attempts at the transformation of everyday life as made manifest in the Events of May 1968. I contend that as a quality of intellectual practice, theoretical inquiry names combining processes at work in the formation of a "consciousness" of the sensible and intelligible, of the basic givenness of the natural and elemental worlds of experience. It is operative in the presentations of the sensible, how sense itself becomes present.

In the next section, my intention is to portray perceptual sense activity as the core of experience and the growth of knowledge, giving a "sense of sense" to contemporary theory, research, and practice. In subsequent chapters I will review some of these practices as they have come to expression in the technosciences. In the following I select aspects of some critical thought, stressing a materially accented phenomenology, starting with particular aspects of Merleau-Ponty, especially his influential turn away from Husserl and all philosophies of consciousness and toward the outlines of a critical ontology. I adopt the premises of what Mark Johnson (2007) and others refer to as "embodied cognition," in depicting what Johnson felicitously calls an "aesthetics of understanding." I draw upon insights from disparate traditions, including those associated with the work of Deleuze and Foucault, in order to regard thinking's emergence. The common question and undercurrent is a consideration of the origins, limits, and horizons of human theoretical practice.

Embodying Thought

A phenomenological reading of theoretical curiosity entails a completing of a circle. It suggests a description of a unity in diversity, of a fullness to experience and thought. The headlamps of intellectual inquiry often light narrow tunnels, mining within a polarized epistemological and ontological compass. Ancient walls still stand in place that divide and abstract theory from practice. Much of our contemporary intellectual, academic, and political traditions sustain a depleted and dualistic rationalism. Even in the purportedly empirical sciences, an incorporeal reason often bifurcates sense perception and knowledge, embodiment and sustained disciplinary practices. Idealist, realist, and naturalist renderings of knowledge alike can ossify inquiry in their claims to objective, atemporal truths. They have lost their ground and constituting origins in perceptual experience and the generativity of sense as the primal source of meaning. As Husserl (1970) claimed in the *Crisis of European Sciences,* intellectual and scientific research programs, "the passionate praxis of inquiry," are marked by a generalized disengagement from lived and social experience (p. 40).

The origins and foundations of inquiries in the sciences, at least since Galileo formulated mathematical models, have been obscured by a technization of thought. The sensory bases for knowledge and inquiry, as in case for geometry, arise from their prescientific experience in our common "lifeworld." Any reclamation of the origins of theoretical praxis, as in Kant, is the "motif of the knower's reflecting upon himself and his knowing life in which all the scientific structures that are valid for him occur purposefully, are stored up as acquisitions, and have become and continue to become freely available" (in Husserl, pp. 97–98).

It is a question of the relation of the sensibly known or knowable and the invisible or supersensory aspects of nature. The given "sense-data" do not provide the schema or explanatory models for the "something-in-general," of laws, principles and theoretical developments in the sciences, but are the first pivot and historical support for any research attempting to move from the practical and perceptual into the theoretical and abstract structures and processes at work in a given field. Husserl, remarking on the lack of interest in these primary relations finds that "there has never been a scientific inquiry into the way in which the lifeworld constantly functions as subsoil, into how its manifold prelogical validities act as grounds for the logical ones, for theoretical truths" (p. 124).

In large measure, Husserl's late work strived ardently to ward off the irrationality he could not avoid in the 1930s with the frontal rise of fascism in general, and a resurgence of what he held to a mystifying metaphysics in his former student, Martin Heidegger, in particular. The primacy of perception as a fundamental immersion or "enworlding," completes a circuit attenuated by the institutionalization of binaries and dualisms. Intellectual works, especially in "the ruins of the university," as Readings (1996) observed, have reproduced these divides, granting privilege and authority to the epistemological over the embodied. A common feature of much traditional pedagogy, especially those preparing technicians and technocrats, is to put an "end to the questioning." For Readings, this is an

> orientation toward autonomy, an assertion that knowledge involves the abandonment of a network of ethical obligations: to have knowledge is to gain a self-sufficient monologic voice.... The authoritative voice of the *magister* rests upon his or her (usually his) privileged relation to the meaning of knowledge (pp. 156–157).

Mental representations, neural networks, schemata, reign supreme. This is also evident in the case of analytic philosophical rigor, as evident in its endless pursuit of its "mind-body and problem" and its variations in the new notions of "extended mind". Purportedly timeless structures or neurobiological processes dominate over appreciation of the empirical, aesthetic, and material contexts of experience. A waking amnesia operates with regard to the question of histories of the ways by which our senses are educated.

I wish to take up this reading by suggesting the value of thinking in and with the currents of theoretical curiosity. In its incalculably diverse appearances it is the usually unnamed experience of researchers and thinkers coming upon open and often unexpected horizons. As a way of giving name and description to some aspects of this experience my effort here is to encourage its robust appreciation in all fields of sustained inquiry. I only wish this work to allow acknowledgement of its active role in regenerating realms of cultural potential. A revalorization of theoretical curiosity across disciplinary limits would mean a new social imaginary. In giving a new figure for this otherwise marginal, or minor, and, admittedly, elusive phenomena, signifies renewed roles and relations for participating in the cultures of natural and the social sciences. In no way is this to claim that theoretical curiosity is an autonomous causal agent. It is one aspect or force shaping and shaped by shared human experience and received knowledge. In all its motivations, including the Baconian dominion over nature in the age of European "discovery," it has always been channeled in concert with other social structures. It remains a floating signifier for the historical apertures for innovation within the parameters of existing research, disciplinary methods, and modes of action and reflection.

In this project my intent is to sketch theoretical curiosity in the spirit of a critical and materialist accented phenomenological inquiry. I want to draw into relief how it is a participant rather than an idling spectator to the constitution of the historical imagination of humanity itself. Never a sideshow for the potential for a radical refiguring of the interests of knowledge, I claim that it is one of the prime elements that color and texture the sustained inquiries and research programs, inventions, and occasional interventions of political, pedagogical, technoscientific, and philosophical disciplines. To attune ourselves to sense, the senses of sense making is to engage in the processes that create structures to our experience. Sense and perceptual primacy are layers, folds in experience; not merely sources, they are also the grounds from which knowledge, concept, theory, objectivity, and language emerge. In granting perception primacy this approach in no way diminishes the place of the conceptual or cognitive aspects of human labor. It is rather to emphatically draw their unity in difference. To do otherwise, would be to reduce theoretical practice to the atoms of the first empiricists.

Perception is the starting point for all our expressions. It is also, sensibly, the point of a continual return for any embodied practice participating and contributing to discovery and invention in all its symbolic, linguistic, and artistic elaborations. For Merleau-Ponty (1997) they are consubstantial, comprised of the same "flesh" in which interiority and exteriority exist in a continuum. Language, conceptual and theoretical work, including that of scientific research, is a "second visibility," another fold in the textures of experience. Language and conceptual representation, as forms of imaging, give expression to their entwinement in a primary perceptual and sensory reciprocity. To turn curiosity toward itself, becoming "reduced," for my project's phenomenological moment, is to become curious about curiosity, a reencounter with a world always in the making.

Enworlding

We are always in the midst of things. There is no place apart. The lifeworld is the originary, relational habitation for experience and experiment alike. Its filaments intertwine, relay, and reprise the historical and material conditions of our coexistence. Cohabitation is always founded, existing on the basis of often unacknowledged, pre-conceptual, "pre-predicative" grounds. These givens set limits to the order and compass of everyday forms, expressions, and configurations of thought, affect, and subjectivity. From a phenomenological perspective, there is no pure outside to existence. At the same time, lived experience is the measure and criterion of authenticity, and, as claimed by pragmatists, always remains open to a continual reflection and radical questioning. The "natural attitude" and the "perceptual faith" discussed in the remarkably suggestive passages in the *Visible and the Invisible* are the everyday experiences and sense world of perceptual activity; they are the unquestioned givens of human sensory capacity. For Husserl as well as for Merleau-Ponty, it refers to the world as an unquestioned, constituted, and unreflected entity. It is the domain of the everyday experience of human subjectivity.

Eugen Fink, who was Husserl's assistant, described the phenomenological method of "reduction" or "bracketing" of the natural attitude as a process intended to temporarily suspend the perceptual and sense-making elements of human experience. These ever-present qualities involve our continually reciprocal relations with the world and with other beings. Commenting on the "natural attitude," he accents an essential "enworlding" that comprises both our presumed ideas of the world and of being, the "setting up of man as a being in the whole of the world." Instead of using Husserl's favored term, the unreflected "natural attitude," he prefers to write about our "captivation by and to the world" (1995, p. li). As a process of disclosing how we are already constituted in a world of perceptual activity and reciprocity, the phenomenological approach is intended to "elucidate the conditions of possibility for that reflective 'gaze' itself" (p. li). This requires describing the open qualities, free variations, and kinds of experiences of inhabiting our place and relations within ordinary, everyday life. For Fink the world is the "universal horizon of being," in which we find ourselves "bewitched" (p. liii).

The natural or "naïve" attitude provides one guiding thread for the phenomenological project. In historicizing our experiences, I want to suggest how it can bear a critical and materialist motivation. In this vein the natural attitude is a common ground for any pedagogical, philosophical, and ethical project whose intent is to historicize the panoply of human experience, practice, and ways of knowing. It is directed to a restless interrogation of the conditions of possibility for the critical generation of the sense we make of our contingency, of human labor and passions. The world as given to immediate experience, the sensory and perceptual manifold encompassing human experience, memory, and imagination, is the wellspring for any form of knowledge we may draw from it. Critical theoretical practice is nurtured from these primal sources to participate in a purposive and unfinalizable questioning of the fonts of experience, knowledge, language, and social practices.

For the interpretations of the phenomenological tradition that inspires this project, my emphasis accents the constitution of the meanings of everyday life, as well as their overlapping leaves, in the sciences, the arts, pedagogy, and social movements. The reciprocal relation between perception and knowledge, of thought and language, is a pivot or "juncture" for all our embodied movements; we can only participate in "making sense" of our collective lives in a world that is constituted in and by perceptual sense itself as our primary activity as sentient beings. Human practices in everyday life are performed in the "natural attitude." They are structured and patterned in the reciprocal copresence and interrelation of historically received and performed knowledges, sense making, and perception. Both the "perceptual faith" of the unreflected life, as well as any inquiry of it, are grounded in the fields of perceptual co-participation, in a shared material world.

Merleau-Ponty's (2010) concern for what he regarded as generative processes of "institution" was motivated by the usurping of the given and unreflected dimensions of our already sociohistorically constituted lives, thoughts, and desires. The sediments of surviving concepts, practices, or introjects are inherited through our diverse structures of socialization. In the contexts of their transmission, particularly in situated pedagogical shaping of the senses, they are subject to continual renewal and regeneration. He offers a number of descriptions of institution quite pertinent to my intent to articulate a critical theoretical practice. This includes the notion that "institution in the strong sense is the symbolic matrix that results in the openness of a field, of a future according to certain dimensions, and from this result we have the possibility of a common adventure and of a history as consciousness. Why does something advent?" (p. 13).

In light of this and questions raised in the last chapter, regarding the forays of a critical curiosity intent on changing life, we can also ask: how do we make history? Theoretical curiosity is a mode of expanding and connecting existing disciplinary knowledges. It is a name for a continuous process of reflecting on the "natural attitude" of our practices, as well as our concepts and uses of knowledge. The reserves of knowledge are vortices, material assemblages of methods, instruments, institutions, and discourses, reciprocally active and shaped within the fields of power in which they and their disciplines are crystallized.

How is it that, like small flames, questions arise from the filaments of our senses? To practice theoretical curiosity is one way to attend to the opening of horizons and the depths of possible horizons within them in the remaking of the world for new purposes and "styles of being." The phenomenological accent of my writing is predicated on the core notion, long argued for by many over the centuries, of our existence as embodied, perceptual beings in constant interrelations within worlds of experience always open to transformation. A constituting, concrete subjectivity forms the moving ground of our experience, knowledge and relations. This stress on embodied perceptual activity entails appreciating our sensory and perceptual endowments as powerful bearers of meanings, structures of shared and negotiated qualities of experience. Perceptual sense is already meaning. Perception and sense are registers for the production of meanings. They birth the full spectrum of the structures of experience operative in everyday activity. Figures, concepts, and affective orientations are contoured through the intersubjective activity within communities of practice as they historically accumulate social and theoretical capital.

The natural attitude, the acceptance of the nature of the world as fixed, unreflected, and unquestioned is the manner in which experience is performed in spectacle of everyday life. It is the quotidian traffic or "chatter" of daily interpersonal responsibility. The natural attitude from which critical reflection emerges is to be found in the repeated patterns of social rituals, their routines, codes, representational order, as they regulate qualitative relations between structure and agency. From the perspective adopted here, its apparent immediacy is actually a very highly mediated reality. For Barabas (2006), any "access to the immediate is anything but immediate" (p. 5). The nature of the immediate, as essentially "visualist" representation, benefits from continual "reduction" and reflection. This is a perspective that regards the sensory field as singularly capacious. Husserl's premises regarding perception gain a "primordial status," for Barabas, in that it opens access to the presence of actual being itself. In exceeding all classical categories and spectra, sense perception, the aesthetic, including Locke's "sense data" and the epistemic, then, this renewal is always already present as "perception comes to exceed the level of empirical perception,—strictly speaking sensory perception... and can encompass intellectual intuition" (p. 11). The "larger sense" of a Husserlian appreciation of perception as presentation of sense, amounts to a "broad or suprasensory concept...In fact, it opens the way to a definition of perception detached from any reference to sensory data" (p. 11). It allows for a rethinking of the very ground of a sensory manifold, of the nature of what constitutes, in Aristotle's terms, a common sense. It suggests a vital revaluation of the commodified relations of lived experience, perception, and thought.

Crisscrossed in layered relations, everyday experience requires the mediation of our senses; they inform, shape, and synthesize activity, thought, and expression. Any inquiry into how the given world and the natural attitude are produced by many folds of sedimented and shifting historical forms of identity, including lived experiences of race, sexuality, kinship and family organization, religion, and economy, must "pass through an inquiry into perception" (p. 11). The "naïve" or natural attitude, however, takes for granted the structure and qualities of perception, knowledge claims, and judgments. Lived experience, for all its apparent significance, obviously cannot stand alone as their final arbiter. Especially in the realm of theoretical inquiry and progress, as in the natural sciences, another complementary (though not "higher") layer dimension is conjoined to the unreflected sensorial impressions of everyday perception.

Experience itself, initially appearing immediate, is the font of the world's upsurge and presentation. The qualitative, sensory world, as the completely grasped totality of possible lifeworlds, is the manifold of potential perspectives or horizons of sense. These horizons of the senses of the world, and the sense of sense, are themselves always open, radial, centrifugal. Perceptual sense, in principle, can produce infinite variations by and for individuals and critical disciplinary communities. There are always horizons within horizons, or as Husserl asserted, grounds within grounds of possible worlds of exploration and understanding. They might emerge in all manner and fashion of thought; in autopoietic, self-reproducing systems, as Luhman (1995) or Maturana and Varela (1980) observe in bioinformatic structures, in any formal instances, as in academic habitus, or in hierarchical, serial, of fractal phenomena, discerned in the natural sciences. They are never closed, sealed off from access to new variations.

John Sallis (2000) emphasizes the significance of the phenomenological method of free variations. This is the practice of theoretical or instrumental adjusting how things come to appearance. This includes altering our bodies' orientation contact with phenomena, a kinetic perceptual reframing of the gestalt structures that Merleau-Ponty adopted from its primarily German school of psychologists. These reorienting ways of perceiving include "lateral" and "peripheral" presentations, of the ways by which they come to presence for us (p. 116). Representations and their instrumentation are forms of capture of a more fugitive presentation. The horizons of presentation are always singular and contingent, open and subject to instantaneous change, as evidenced perhaps in studies of quanta. The phenomenological question pertinent to theoretical inquiry here is how the appearance or "modes of givenness" of our world comes to be knowable, how sense is historically configured in our practices, ways of being, and in the prospect of new ways of becoming.

Perception is the bearer, not the mere receptor, of historical and individual knowledge. Perceptual sense needs to be freed of a persistent legacy that delegates it as sense-data, qualia, atomistic sense-impressions, or the ephemeral *phantasmata* of a usually ocular preoccupation. Perception has routinely been distanced and devalued in realist, empirical, and idealist conceptions alike from the traditions of thought initiated by Empodcles's "effluences" and "pores," Aristotle's "rays" or Neoplatonist and empiricist accounts of atoms or sensations. Thinking perception in realist or empiricist terms reduces the realm of the sensory as one comprised only by and for entirely external goads, stimuli, data through management by a rationalist intellectual will to power. Perceptual richness is robbed by a supervening, self-authorizing rationality acting in essential autonomy from the material world.

The phenomenological project as reworked by Merleau-Ponty and all those whose concern is a non-dualistic embodied way of participating in the currents of contemporary life. It places perception as the primary matrix for human experience and knowledge. It resituates the otherwise entirely valorized epistemological fascination with "theories of the mind" and language. Merleau-Ponty works against the classical and modernist grain by which the aesthetic dimension of experience was opposed to permanent structures and forms of intellectual knowledge and truth. The virulence of a dogmatic rationalism, metaphysical idealism or "intellectualism," the aesthetic and perceptual have been derogated as an inhuman purity of thought. These traditions have partitioned life from thought, forming a primary fault line that delegated living bodies, corporeality, the carnal, and intersubjective material worlds to inferior states of being and knowing. Perception was subjected to the timeless forms, rationality ascribed only to the soul or intellectual faculties. Merleau-Ponty is only one voice that would overturn the dominance of the epistemic, of an "autonomous" mind over the perceptual, lived body. He wanted to make modern thought acknowledge this immense epistemic detour. For Husserl, the founder of a "science" of phenomenology, it is a necessary homecoming to an originary, Hericltean flux. It is a return from fixed logicist, positivist, and dualist entrapments. For Merleau-Ponty (2002), shedding much of the idealism and rationality of his predecessor, the sensible is paramount and crux, the "peculiar product of an attitude of curiosity or observation" (p. 263). The phenomenological interrogation is kindred with a radical imagination, venturing to disclose dimensions, turns, and lines of flight across and within the horizons of lived experience.

Chiasms

The core dualism's founding Western culture and thought have met continual resistance. Their perennial attempts to find commonality resonate strongly in modernity. Beginning perhaps with Nietzsche's strategic insurgence into the legacy of Platonism, it has continued with Husserl, Bergson, and with James, Peirce, and Dewey in the American pragmatist tradition, and with the gestalt psychologists and the phenomenological and existential Marxists. An innovative approach to resolving and moving beyond dualistic, abstracting practices is evident in the work of thinkers like Mark Johnson whose concern for an aesthetic understanding of cognition continues an abiding interest in embodied practices Research in the area of activity theory, deriving from the pioneering work of Vygotsky, Luria, Zinchenko, Elkonin, and others, is also founded in situated and material contexts of knowledge and relation.

For this project the question of embodiment and a unity in difference is very germane for thinking the nature and development of sustained intellectual projects. I read a pivotal place for a materialist appreciation of thought in the phenomenological researches starting with Husserl, and continuing in various directions and interpretations taken by Heidegger, Merleau-Ponty, and others outside the confines of a self-consciously phenomenological tradition. Merleau-Ponty thinks of an immediacy or simultaneity of levels of experience, both spatial and temporal, that bridge the differentiations in lived experience, perception, and concepts. An enveloping and elemental "flesh" is interwoven across the sensible and the intelligible. Acting as a double of sense and thought, the flesh acts as the flashpoint or arc of elemental reciprocal relations inherent to the materially sensible world the phenomenal body inhabits. It is a medium and matrix into which and from which all integrations and differentiations arise, including the visible and the invisible, the sensible and the intelligible, the perceptual and the conceptual.

Merleau-Ponty's notion of the flesh reorients our regard for the nature of embodied experience. It revalorizes the bodily and sensible in its knotting with cognitive processes. Once again, this is to emphasize that the fields of the intelligible are not permanently isolable from the sensible. What once stood as the paradigmatic ideal within the Platonic Academy—and for the moderns, in the development of scientific concepts, abstract thought, and verifiable propositional knowledge—is brought to their origins and put to work in the lifeworlds of the contemporary city. These were questions first debated by the Epicureans and Stoics regarding the relation between sense images, phantasmata, *aiestheis*, *eidos,* and *logos*. Aristotle's common sensibles, forms, or iconic objects of perception, not limited to any one sense, come to appear in fields of already constituted meaning. Perception is always a relational and embodied reciprocity between and within the lived body and its lifeworld. Merleau-Ponty (2002) asserts that "every perception takes place in an atmosphere of generality" (p. 250). While "all knowledge takes place within the horizons opened up by perception" (p. 241), creating and expressing a "certain rhythm of existence" (p. 246), they are also partial, as hidden as they are revealing. Sense gives focus to the intentional perceptual objects it is motivated to pursue. An inexhaustible totality always surrounds any given field which "spread a horizon of things which are seen, or which are even invisible" (p. 251).

The flesh is metaphorically described as an interlacing or interweaving, an open thinking and interrogation of Being. It is the repository and reserve of perceptual possibility, the generativity of thought in a matrix of affect and the five senses. The flesh is a revitalized consideration of the Presocratics' elements. It is a material lattice. This confluence of the aesthetic and the intellectual or theoretical is a dynamic, living unity in difference, as in the membranes comprising the porous textures of a leaf. I emphasize this point as it marks a fundamental departure from all forms of epistemological duality. It is, one, striking way of speaking what is unthought in all variants of a positivist objectivist rationality that excludes the place of the observer as an embodied, relational subject. As a critique it also responds to idealist and empiricist models. Merleau-Ponty is in continual confrontation with the prevailing arguments of those who make hollow epistemological claims for the faculties of an autonomous mind as well as for those realists and empiricists who separate perception and meaning from embodiment. His adversaries included all those who practice some form of reflection theories of meaning and the making of sense. He engaged traditions, whether in Cartesianism, or in Husserl's transcendental reduction as a rationalist philosophy of consciousness, as well as then presiding forms of intuitionism as advanced by Bergson. It is a theoretical departure that begins by proposing instead of the tenets of these thinkers and their practices, an originary ontological ground. Merleau-Ponty critically interrogated his contemporaries as well as his own prior subject-centered thinking. He continually attempted to sketch out an "a-philosophy," a monist and critical ontology. His reworking of traditions offered a renewed way of appreciating our embodied place in the world. It is a wellspring that draws from the primary, dynamic unity in difference of our being. His work is a rich resource for any pedagogical projects trying to join and fuse the splits in the social and human sciences.

Merleau-Ponty's "flesh" opens thinking to a sense of the doubled nature of our embodiment. In contrasting the lived, phenomenal body from an objectified, clinical body for inspection, he stresses the importance of its movement inside its environment. The lived body inhabits the meanings and the places of its lifeworld. In a sense later articulated by both Deleuze (1988a) and Foucault (1977) there is a manner of describing the phenomenal body and experience as always simultaneously interior and exterior. There is in a phrase always, an "inside of the outside and the outside of the inside" in the relation between the sensible and intelligible, the visible and the invisible. He borrows Husserl's image of an individual touching one hand with the other, their folding over mixing the sensed with sensibility, the "outside" hand making contact with the inside one in a reversible reciprocity that becomes emblematic for the nature of our perceptual integration within a world that can no longer be posed as exterior. The horizons of our sense perception may always be partial, but, like the gestalts that inform much of his thinking here, are very finely configured within those living fields in which we constitute and perform our kinetic practices. It is a way of describing experience without borders between the spatial and the temporal, nature and "instituted" practices including the "higher mental functions" of language, prediction, and theory. It affirms the senses in the making of language, concepts, and all forms of abstraction.

For the pragmatist tradition discussed in Chapter 3, and for Dewey in particular, we are always engaged as participants in situations. It is the principle of continuity across the spectrum of human practice, across which the sensory or aesthetic infuses cognition, theory, or concept. This is also the world of qualia for cognitive science. The qualitative, lived experience of embodied cognition can never be meaningfully reduced to abstractions. Qualia, as Mark Johnson so ably describes, are "felt qualities, like the blueness of a blue sky or the silkiness of a silk dress or the smell of summer lilacs. The problem is that qualia cannot be reduced to conceptual structures or to functional states of an organism" (p. 70).

This is also the canvass on which the phenomenological forays of Merleau-Ponty draw its images of thought. Concepts, linguistic propositions, and all realist representational paradigms derived aspects form a primary, perceptual situatedness. It is an immanent, embodied crisscrossing, a "chiasm" of the bodies in and of the world. This "flesh" of the world excludes none of the traditional dichotomies of practical reason, including that between the sensible and the intelligible, *aistheis* and *episteme*, thought and extension, or the visible and the invisible. The perceptual image exists in a circularity or circuit of experience, into which intellectual or theoretical tenets prosper. It is a fold or membrane for Merleau-Ponty (1997), a "second visibility":

> With the first vision, the first contact, the first pleasure, there is initiation, that is, not the positing of a content, but the opening of a dimension that can never again be closed, the establishment of a level in terms of which every other experience will henceforth be situated. The idea is at this level, this dimension (p. 151).

Revisiting the Presocratic pursuit of the nature of things discussed in the opening chapter, Merleau-Ponty's fragmentary and eloquent depiction of the "flesh" is crucial for my abiding concern for the place and play of what Aristotle famously proposed as a primary desire and pursuit of knowledge. In the lectures the Stagarite performed at his Lyceum, only four senses were detailed. They were said to correspond with the four elements. Sound was the analogue or correlate to air, water with vision, fire with smell, and earth with touch, the primary and generative sense. Taste was relegated to a specific form of touch. He contributed to the long-standing interest in whether or not an overarching and inclusive "common sense" existed. This *sensus communis* worked as the messenger and mediator between the other four.

While the "natural doctors" advocated for a distinctive and primary founding substance, whether designated as air, earth, water, fire, or the boundless of Anaximander, Merleau-Ponty wants to seize the notion of an elemental and unifying quality for all lived sensibility. The "flesh" is his common sense. The flesh is a way of using language to describe something it participates in and derives from itself. The "flesh" is neither the physical skin of sentient life, nor the corpuscular, venous, and neurological. It is at once the sensuous contact and perceptual copresence in an environmental situation, milieu, or culture primary to the sustenance, and very existence, of all forms of life, from one-celled organisms through the evolutionary links with mammals and primates.

To provide a way to rethink the sensible and the differentiations that are discovered and explored both in everyday experience and in reflective inquiry, Merleau-Ponty (1997) describes the flesh as

> not matter, is not mind, is not substance. To designate it, we should need the old term "element," in the sense that it was used to speak of water, air, earth, and fire, that is, in the sense of a general thing, midway between the spatiotemporal individual and the idea, a sort of incarnate principle that brings a style of being wherever there is a fragment of being. The flesh is in this sense an "element" of Being. Not a fact or a sum of facts, and yet adherent to location and to the now. Much more: the inauguration of the where and when, the possibility and exigency for the fact; in a word: facticity, what makes the fact be a fact (pp. 139–140).

This project pivots on the premise that a genuinely theoretical curiosity is a material, embodied practice. As a practice of the mind and mental constructs, it most certainly generates conceptual and abstract forms of thinking, design, and the building of models. The link that I wish to stress is one between the practice of thought and perception. This unity of difference in our sensibility is the generator of the meanings by which the world may be given shape in our projects and disciplines. It is the double bond between the horizons of sense and thought. Sense continually presents, rather than represents, meaning, as the "chiasm" or "interleaving" juncture that Merleau-Ponty describes when he presents the "flesh" as the connecting tissue of world and being. As an embodied practice, theoretical curiosity retrieves and returns, in cycles of reflection and action, to the sensory richness of our perceptual origins.

Our thoughts are always directed. Intentionally drawn and responsive to things, theoretical forays of "thought" or "mind" are at once material and embedded in our everyday lives. Thinking is a moving and orienting toward and in relation to "things themselves" in all and any of their angles, horizons, depths, and variation of "untouched" or hidden dimensions. Sense is directed, intentional, and situated in the indissolvable immersion of bodies in the world. Thinking in this sense is motivated by and with actual and virtual objects of inquiry. The critical aspect of this desire for knowledge is initiated by the urge and possibility of finding new faces to the "things" in the discrete worlds of our practices, to the studies to which we attend and to which we would become more finely attuned. It is this reciprocal making of newly viable pedagogical and cultural environments that is of most concern for me in thinking the value of inquiries that, at first glance, may not seem to offer a plausible joining together of otherwise-polarized spheres of human experience. This is especially germane to institutional and corporate models, and the depleted relation between perceptual acuity and the place of informed reason and rationality. It is always a question of the degree to which these links remain open, conscious to those who practice them. This is especially pertinent for the designers of new artifacts and technics in technoscience projects as discussed in subsequent chapters.

Throughout all of our contemporary theoretical projects there is a potential movement outside representation, monadic interiors, or the boundless exterior for Promethean projects. In making actual the potentials of human praxis in the material world, theoretical forays come carrying considerable ethical cargo. As, for instance, in experiments and innovations in urban architecture, they can bring real, experiential shape to our spatial and temporal horizons. These horizons are implicit in the mission statements, ideals, covenants, constitutions, contracts, and aspirations of the disciplines of knowledge. They are the forms of capture and use to which the labors of thought are marshaled and mobilized. They are signs of our collective, often unrealized intentions.

Elemental in its sensory indeterminacy, this "passion of the mind" is a passion of bodies in their charged, often volatile, interchanges. The critical questioning of relations, things, and experiences that I profile here is not a guise of theory itself. It is not necessarily a facet of a "theoretical attitude" that could readily be isolated from the entanglements of historical forces, techniques, and material practices of everyday life and its sustenance. I wish to claim that what has been named theoretical curiosity may be better appreciated as *the desire of theory* itself in its generation of research projects. It is neither the individual psychosexual fancy of popular culture nor comprised by the empirical generations, predictive power, or data and evidence sought and generated by theoretical work or practice. This elemental, affective spur to thought and activity is, paradoxically, neither limited nor limitless. It cannot meaningfully be reduced to or by the visualist gaze of classical and modernist theoretical speculation. It throws down a gauntlet to what Kant found wanting in all the "artists of reason" manifest in the speculations of geometers and mathematicians. I believe it can intercept the formulation of any Promethean and totalizing "theory of everything." It defies the absolutes of an objectifying, scientistic gaze. It suggests the immensity of sense and the qualities of all modes of perception in and within what Merleau-Ponty heard in a contextually engaged "singing of the world." I place emphasis on its affective intensity in sustained and collective inquiry. It is an affect that fuels the pursuit of discovery and adventure, of exploits and interventions, clinical trials and endless empirical researches in even constrained research program parameters.

Theoretical desire always takes on the color and tones of its generation. This was the case even in the dematerialized mathematical model of early modernism illumined by its "natural light." In the spectral perfection of a Cartesian mathematical model, theoretical objects exist without extension; in the *res cogitans*, a virtual infinity of geometric functions, ratios, and degrees perform without subjects, movement sound, color, or taste. It is the translucence of a truth and perfectibility accessible only to mathematical intuition. Its natural light is its truth and guarantor. The colors of a phenomenology of experience, at least within Husserl's work, do not mix into the full palette of sensuous perception. The care for "rays" of attention, the "intuitions" of the essential, and the use of the term *eidos* in his work (as well as Heidegger, discussed in the next section) still center on images of light, "clearings," "showing," and "shining." The visual is the imposition of a spectator, of a nameless objectifier and experimenter, one pertinent to my use of Heidegger and Janicaud's critique of technology in the next chapter. It haunts much of contemporary representational theories' implicit operating principles. The objectifier's methods are rarely transparent themselves. When they are scrutinized, they can function as self-authorizing and atemporal means to disclosing the phenomenal world. Within the original Husserlian phenomenological method, too, particularly in the "eidetic reduction," an implicit optic figures all constituting intentionality. A visualist will to representation animates our ability to make things appear to us as objects of possible knowledge.

For Husserl, this intellectualist or "noetic" process comprises a primordial cognitive power. Its intentionality is still the prism for a transcendental subject that generates and makes present the objectifiable world of perception. All possible knowledge, of both its spatial and temporal dimensions—mathematical and experiential, virtual and perceptual—can be generated in its limitless variations. The hegemony of the visual remains an illuminating blindspot. It keeps us from an expanding sense of our embodied knowledges in what Don Ihde (2007) asserts is the fulfillment of perception a revitalized, elemental phenomenology seeks. Through the directedness of intentionality toward and into and certain horizons focused aspects of perceptual experience become theoretical and problematic epistemic "objects." Some aspects are disclosed and present to inquiring perceivers while an incalculable quality and quantity remain peripheral, hidden, and unexplored in the interests of delimiting a discrete and practical research program. It is in a continual redirecting of attention and in seeking unexamined perspectives that a synthesis of sense knowledge can renew our bearings and open us to the future. For Ihde, this "opening" or beginning presents us with a "rediscovery of the richness of experience and its structures is a discovery of the essential embedment of experience in historicality." The essentially "global" nature of experience and perception in and of a world we are immersed in offers up awareness of the "polymorphous flexibility of human being" (p. 20).

Returning to beginnings and becoming beginners ourselves is also to ply time. Giving shape to time's experiential unfolding, this is a freshness in always beginning again that phenomenology invites. In the context of the phenomenological approaches discussed here, it is the practice of free variations and the reversibility of perception, one that is not limited to a single sense but to their concert in attending to the possible. One accent can be placed on temporality itself. Among the currents that carry sustained inquiry, there is often a temporary suspension of will, or the arrest of grand narratives, plans of action, teleology, or systematic investigation. Kindred to wonder, astonishment, and awe, as in Descartes's anatomy of the "passions of the soul," curiosity may also be a displacing detour within the experience of temporality itself. Like stupefaction, it interrupts analytic rationality, which Avital Ronell (2002), commenting on a primary philosophical questioning "throws one off; it involves a stepping back from immediacy" (p. 111). The passion for knowledge often willfully violates the spatial and temporal borders of an era, its "faltering step" calling into question its existing conceptual cartography. The sense of time can partake the qualities of lived durations, not mechanical chronology or linear diachrony, in which affect and relations are given contour. As in many pre-industrial and nomadic groups, a rhythmic or "vertical" time can be set into motion. A rhythmic sense gives shape to the events of individual and can form the experiential structures of individual and collective activity. It is a repartitioning of the temporal, a handcrafting of finitude.

An elemental, materialist curiosity takes on the freight of the historical persona that puts it into service. Without essence, it appears in distinct epistemic regimes fashioned within their sanctions. Despite the elusiveness of this desire or "drive," I contend that it forms the energy of sustained thinking and care, a collective and individual "wanting of the mind." As a transhistorical potentiality, realized, censored, or mediated, the desires of theory ally with the imagination and the acuity of our sensory manifold. Unlike the instincts or any claim to a purely neurobiological inlay, it is composed of human perceptual attunements when allowed to flourish and find expression. Through a "looping" of reflection on sense, or what Ihde calls the "experience of experience," this attending engages historical and subjective memory, concepts, and the sedimented images and impressions of the education of the senses. As a "stepping back" and taking distance from experience and relations, reflection emerges from the attention to things that curiosity presents. For Ihde, reflection constitutes our awareness of being in the world. It is "a special mode of the relation between my being in the world" (p. 37). Any "self" awareness dawns with primary perceptual, sense experience. This knowledge or sense of being is mediated and indirect in that it is always made possible as a reflection or response to the surrounding world. I am not claiming that this desire, force, impulse, or drive is the root of human experience and knowledge. To reduce this elemental pursuit and desire for knowledge to the caprices of a straying, primarily visual perception, or idealized into the sterility of the epistemic, marks a historical aberration for the experience and experiments of all thinking bodies. As Dewey advocated, no pedagogy, in either the natural or social science, should ever cut off the experiential from experimental learning. Disciplinary knowledges are stealers of a fire. They can traduce the existential openness of human becoming.

Sense and Sight

This section will further explore the enduring question of the visual in its relation to theoretical curiosity's pursuits. Though certainly not an uncommon intellectual and academic concern, my purpose is again to place accent on how the "regimes" of the visual and curiosity have historically been paired together. From an idealist standpoint, their relation is inherent, an instinctual drive or appetite.

Sight remains central to the purposes and projects of Merleau-Ponty and Heidegger. Merleau-Ponty's unfinished last book, the *Visible and the Invisible,* appears to have begun the project of formulating a critical ontology. The notions of the flesh and the chiasm were tentative explorations for an "a-philosophy" of experience and everyday life and thought. It would be one founded on an openness of perception and sensibility. It is one in which interiority and exteriority, my body and the world, are inseparably intertwined. Without flattening into a homogenous equivalence, there exists an ever-developing attunement between perception, of things, and the world of embodied life. Sight and visibility are inherent to any phenomenological project that, though traversing language, moves beyond the object-subject, binary grid taken for granted as we inhabit the natural attitude. Sight as the primary sense takes central stage within the perceptual nexus of being and becoming. He asserts that "our world is principally and essentially visual; one would not make a world out of scents and sounds" (p. 115).

His attention to sight was not intended to support a representational order centered on the ocular in all its powers of seduction. Merleau-Ponty's attention toward the visual and vision was directed toward its function as a primary entry into the singularity of things. In distinction from the other senses with which it co-responds, "the 'visual quale' gives me, and is alone in doing so, the presence of what is not me, of what is simply and fully" (1964, p. 187). There is a lack of a sense of the exterior in the other senses. Hearing seems to occur in the ear, touching "at the tips of the fingers." Instead of an outside that situates the powers of sensibility, we might remain oblivious to the situated place of all or any experienced perception, including its decisive expressions in language, art or theory. Barabas (2004) explains that "by its very exercise, vision ignores itself qua sensible experience; the self-forgetfulness inherent to the natural attitude in achieved in vision" (p. 149). This is the basis for tenets of some realists, providing an illusion that objects always exist in themselves, outside of the variations of instrumental, technical and experiential perception. His target here is not only this "realist illusion" but also the "intellectualist" or idealist error with regard to vision.

Merleau-Ponty's appreciation of vision is of a visuality "pregnant" with the invisible. It is an argument with enduring Cartesian epistemologies, with the eye of the mind. His writing on this theme refutes the premises of an intellectual or intuitive vision. I have noted its persistence and importance to thought since the classical era of Greek theory. Intellectual vision is rootless, disembodied. It does not reflect on itself. In its self-constitution, the radical doubt of the mind and its method is illuminated by a "natural light." As Barabas cogently explicates, "under the pretext that there is no vision without thought, Descartes concludes that is sufficient to think in order to see. Vision is reconceived not as a certain relation to the things themselves but as thought brought into action by signs in the body" (2004, p. 150). The intellectual vision is a play of light that floats transparently within the abstract space and dualism of extended, thinking substance. The ties between theory and speculation, sight and the question of theoretical "seeing" of the unseen, stay core knots for contemporary practices to unravel.

A play of light has figured a visual optic for theoretical curiosity. It is an ancient and continuing source and means for exploration, discovery, and explanation of phenomena. The visual has figured as the optic for genuine being from Plato's blind and vagrant cave dwellers through to contemporary technoscience's scanners and imaging instruments. Heterogeneous cultures have shared some core values emanating from the radiance that Descartes, for instance, hailed as the "natural light." It arises across regimes of thought and practice; it is evident in questions asked about the knowledge gained from the visible and sensible, first posed by the Presocratic "doctors," through Cusa's mirrors, Roger Bacon's images of the rainbow and droplets of water on a glass, into Galileo, Copernicus, and Spinoza's age of lenses, and into the closed spectacle of late capital. Plays of light signify how frontiers are crossed; the detection of subatomic particles or background radiation from the origins of the physical universe participates in an ongoing Promethean optical mapping and representation of what appears to exist beyond natural human sensory experience. The optical in all its technical mediations through scanners and imaging devices, appears to hold precedence of our sense of knowing, recording, and documentation of evidence. It is ubiquitous in contemporary laboratory and research practice, evident in the refracting, reflecting, scanning lenses and mirrors of astrophysics and the fiber optic, bandwidths and visual instrumentation at nanoscales. Hubble and its successors, as evident in the European Space Agency's Planck Space Orbiter and NASA's WIPP, follow a history of exponential detection and mapping, repeating Hubble's doubling of the capacity of Hale's lenses in the early twentieth century.

Heidegger (1962) discusses the primacy of the visual in the Western philosophical tradition. In a section of *Being and Time*, entitled "Curiosity," he indicts nothing less than the whole trajectory of Western metaphysical and philosophical discourse as an evasion and fleeing from a genuine pursuit of truth. Citing Augustine's famous condemnation of this "desire of the eye," Heidegger asserts that philosophy—both classical and modern, from Parmenides to Hegel—has only practiced an ocular surveying of things. Heidegger refigured Aristotle's famous statement as a "seeing, a desire to see," in his translation from the usual "all human beings by nature desire to know" to "the care for seeing is essential to man's being" (p. 215). Theory has the world in its sights. In section 36, Heidegger considers the central role vision has served in framing Western thought. Following Aristotle again, he claimed that "originary and genuine truth lies in pure beholding (or intuiting). This thesis has remained the foundation of western philosophy ever since" (p. 215).

Curiosity is delegated to a mere "seeing" of things pertaining to the lived experience of the mundane and everyday. The philosophical traditions and doctrines emerging from the classical world, for Heidegger, regard only a world of dispersion, of superficial concern. A lack of directedness effaces an originary wonder, *thaumatzein*—as propounded by Aristotle's claim—which from his *Metaphysics* on is the prime motive of all human seeking for knowledge, philosophical, scientific, or otherwise. Sighting shifts its concern or caring. McNeill (1999) reads this section as one attempt by the master of Marburg to make an apparent correspondence, analogue, or alignment between the philosophical desire to see and that of curiosity. Asking a question at the heart of my project regarding theoretical inquiry, McNeill wants to know whether or not

> Heidegger is implying that the whole of western philosophy can somehow be reduced to curiosity? It is difficult to avoid this impression, notwithstanding the assertion that curiosity does not seek to understand what is seen, but seeks "only to see," or "only to see and to have seen" (p. 3).

Sense and theoretical sightings become partners in a continuum. The vast edifice of Western hierarchies and dualities is made ambiguous.

McNeill is an astute reader of the texts in which the concept of the *augenblick*, the glance, glimpse, or "blink of the eye" is pursued (as in its pertinence to Lukacs and revolutionary social theory, discussed in Chapter 4). He asks, "is it not quite remarkable that he should here locate a discussion of this tradition and its emphasis upon seeing within a section that bears the simple title 'curiosity'? (Die Neugier)" (p. 3). The "blink of the eye" is a philosophical and aesthetic motif originating in German romanticism with Schelling, and has been applied to diverse contexts by Kierkegaard, Nietzsche, Jaspers, and the critical theorists associated with the Frankfurt School. It suggests an essential temporal bridge, opening, and relation between theory and the practical world of decision, of *phronesis* and the productive, artisanal world of *techne*. McNeill, following Aristotle's assertion of an originary praxis, points out that the "blink of the eye"

> is not simply opposed or in contrast to any of the traditional polar terms, including importantly theory and practice, theoria and poiesis, episteme and techne, or the intelligible and the sensible. Rather, it puts all such oppositions into question, unsettling them as it acknowledges their critical importance. It is itself the site of *krinein*, of finitude and of the possibility of critique (p. xi).

In this incisive reading, an entire tradition is at stake. The aims of knowledge and its very nature are the subject of an inquiry that contests the foundations of traditional and classical thought.

The contrast between an ordinary seeing, as curiosity, and philosophical sight can be reexamined as expressions and instances of a theoretical attitude intrinsic to human experience. Ordinary or everyday seeing is relegated, by Heidegger, to a mere "desire to see" and to "merely have seen" objects and beings. Following the Patristic tradition, and Augustine's sanguine invective, Heidegger initially repeats the hoary condemnations of curiosity as an excess of the eyes. It is once again regarded as a perennially entropic tendency to dispersion. It is the marker for a lesser order of concern, a lack in our essential attention to the things and events in and of our presence in the world. He identifies ordinary sight with *nuegier*, novelty, with the distraction found in magazines and amusements. Curiosity in the everyday then becomes associated with the decentering of our place in relation to being and becoming, as it does not "tarry" with its objects. For Heidegger, it does not seek to know or to intervene into ways of the world and our possibility for either transforming daily experience or understanding the truths of our coming to be present. This analysis will be useful in my discussion of the place of scientific and technocultural inquiry in Chapters 6 and 7. He considers it as a flight from the openings and decisions that a transformative experience, of a philosophical sight might proffer. This is the moment of the *augenblick*, the glance or blinking of the eye, that is nothing less than an apotheosis.

Philosophical sight as a purely theoretical attitude glances toward the nonsensory in order to seek the truths hidden in the "look" and appearance of things. It is the sighting of the *eidos*, the ideas that, once glimpsed, present the real and permanent forms of cognitive knowledge eclipsed by our distracted attention. This is the place of the pure intellectual realm of *noos* and the *episteme* of the ancients discussed in my first chapter. Theory as a desire to see refers traditionally and here again, to this nonsensuous realm. The philosophical desire is a sight that stays present with and in the presence of its objects. It is directed and participates in a temporality that allows for the making of the decisions presented in those contingent "moments" of the *augenblick* for the shaping the color and quality of existing and future perception and experience. For McNeill a critical ambiguity or equivocation may be at work here. Curiosity is to be regarded as constituted in relation to an inherent and a priori theoretical attitude within the human. What is worthy of our attention is that "curiosity and the desire to see, would have to be understood as the grounds of the a priori constitution of theoria, as its own intrinsic 'condition of possibility' and not vice versa" (p. 5). Curiosity is relegated to being a minor working partner and counterpart for a more robust theoretical sight. What I find especially intriguing here in Heidegger's constant intrigue with contrasting ordinary and philosophical seeing is the primacy granted to a perceptual appearing of the world in a curiosity that, in its looking, spurs all philosophical desire. Critically, these sections of *Being and Time* may pose this purely philosophical, or speculative gaze in continuity with its perceptual counterpart:

> this philosophical desire may be nothing other than a reflection, or the repetition on another level of the everyday tendency of it (McNeill, p. 5).

In his influential etymological dissection of Anaximander's only extant fragment, Heidegger (2002) attempts to clarify the possible meanings of key terms. His ardent hermeneutic disclosing spurs his concern for translating and restoring the fragment's pertinence to a modern world abandoned from a primordial relation to being and dominated by the homogenizing powers of technological control. In this reading, the phrase commonly translated as "the things," τα οντα, that have their origin and "must pass away according to necessity" through the rule and justice of time, becomes a decisively inclusive term for the nature of being itself. It gathers together both natural and human phenomena and effort, techne and praxis. In this lecture this term is recovered for the renewal of ontology Heidegger pursued. The phrase became a centerpiece, displaying a surpassing beyond all the interest in the "things" and entities of nature, of φυσις, a primary topic and concern of all the Presocratics beginning perhaps with Anaximander himself. For the modern ontologist, it brooked a primal river of experience and temporality, motioning toward inquiry into the wonder of being and becoming. It is only with the codification of being into the straits of intelligibility and sensibility, of essence and image, conducted through the lasting authority of the Academics, especially Plato and Aristotle, that the more capacious theoretical sight of the Miletians is lost. If accurate, this rendering is a significant act influencing the textual and philosophic reception of ancient thought. Accepting it would alter the use of the basic terms, tropes, and figures utilized for conceptualizing the nature of existence in the Western intellectual tradition. Much more than eliding a connotative difference in translation, for Heidegger, it amounted to a misrecognition of the work of the Presocratics in a metaphysical tradition that came to conceal an originary relation to being and becoming.

What is at stake here is a closing up of the aperture of wonder in early thought. It is comprised by an elementary surrender, by restrictions on the commingling of human and natural phenomena. It is the beginning of the intellectualization of experience. It provides one unattended structural schema and frame for the investigation of the nature of all things across the emergence of the division of the natural and human domains or sciences, and would usher in the disciplinary boundaries and curricula of the Western scholastic and intellectual traditions. The dominant ethos of practices and reflection on the nature of practice outside of methodology is in question. The epistemic regime we still work within remains framed by an imageless image of the practice of theory. Despite the inflamed sophistry at the heart of his own attempts at university reform in 1930s Germany, Heidegger (1962) claims that

> if the way we normally think within a range of disciplines (such as physics, ethics, philosophy of law, biology, psychology) has no place here—if boundaries between these subjects are lacking- then there is no possibility of trespass or the unjustified transfer of notions from one area to another.

He perceives a concentration and caring for a coming to presence of being in the ancients. This care is a converging of the mind's inquiries prior to its subsequent stylization and specializations with their attendant alliances, rivalries, and the institutionalized divisions in the architecture of knowledge.

A New Materialism

Questions are a fluid medium of thinking. Germinal and immanent to thought, critical questioning as a disciplined practice intervenes within the movement and volatile matrices of thought. In suspending any unreflective practice, a critical theoretical curiosity that confronts prevailing currents of thought, including their implicit epistemological and ontological grounds, also succors the productive energies internal to thinking. I wish to sketch here ways of thinking about thinking's own productive abundances, what I sense to be the unpredictably systemic qualities of the curious circuits of thought and inquiry.

I wish here to follow Merleau-Ponty and others outside of phenomenology, especially Gilles Deleuze (1988b, 1990), in pioneering a "new materialism." To begin to think the theme of the materiality, I want to show, how, among a small and subjective selection of modern contexts, theoretical curiosity is continually rooted in the perceptual engagement with the contingent worlds of its practitioners. The "genetic" priority of the senses is historically conditioned. The purpose of mentioning earlier Husserl's critique of the objectivism of the natural sciences and its paradigms was to bring into open discussion their methods, their instrumentality, and their epistemological objects. This is a critique motivated by a desire to resist the formalized dualities still at work across disciplines, including genomics, astrobiology, and artificial intelligence, that I profile in subsequent chapters in order to raise questions a critically informed phenomenological approach can bring to bear on them. These include the issues of the division between their material practices, their tools and methods, and their material practices, including the structure of the labor and financing of research initiatives in their institutional and commercial support. It is to interrogate the nature of what Husserl would call into question in asking after their residence in or abandonment of a "lifeworld."

Jettisoning dualities between epistemic and perceptual aspects of any practice is an utopian project. It is itself exemplifies a critical theoretical interest. It is part of the unfinished project of humanity if it is to continue in and on the material earth. It calls for a *critical ontology*. The notion of a new materialism is a bridge toward thinking futurity. A materialism generated by the work of a critically informed theoretical curiosity is not bound by the natural attitude and the "sediments" of thought practices of prior models, paradigms, and sacred cows; it is not indifferent, nor is it speculative. My partisan advocacy is directed toward forms of theoretical practice that grant an inherent openness to things, of revitalizing our relations and identities to and with nature and animality. I share with many others a passionate interest in effects of our theoretical ventures on the quality of the lives and aspirations of succeeding generations.

A *critical ontology* opens the closed circle of questions of epistemology, knowledge claims, verification, and adequacy that arise from a quantitative and calculating rationality. From the perspective of this inquiry into the motives within the currents of contemporary research a critical ontology is to engage and intervene in a community of ethical practices and projects. It is of no use to proclaim the need for a critical practice, on an ontology of becoming and embodiment, without displacing the speculative or contemplative theoreticist orientation of liberal and social democratic politics. Ontology is merely a placeholder name, a static state of being and making universal, of what is more likely to be an unreflected historical representation of conditions by which life is lived and experiences are normalized and configured.

Ethical practices pertain to the examined, reciprocal relations between and within research communities—in a constant vigilance toward sustainable, ecological inquiry for all situated participants. It is to break free of using and being used by the coin of the speculative realm. Critical curiosity cannot pay a debt or offer change any longer in the currency of a humanism caught in its dualist cul de sac; embodied reason, the realization of potentials in the liberation of our worldly minds in our bodies of earth, will refuse to be trapped beneath an abstracting instrumental rationality. It is a resistance to the neutering of the collective imagination by monetary relations—in the surplus value of human labor and the global trafficking of its bodies. Humanism, despite its technological descendants in replications, automata, and scanning representations of "posthuman" prodigies, remains anthropomorphic. It sustains an enduring idealism that seeks to find closure and finalize theoretical curiosity. In the spirit of system (like the movement of thought that the *Encyclopedists* and Diderot opposed), this is the dead end of a search for permanent processes, structures, and explanatory models—a "cosmopolitics" that could provide nothing less than the operative principles that generate life forms and "consciousness."

In the often-luminous sketches in *The Visible and the Invisible*, Merleau-Ponty attempted to leap over founding dualisms and the split world. He describes crossing points, pivots, junctures, and layered or folded leaves of thought, sense, and perception. Our languages, even those encoded in mathematical or notational, machinic or cryptographic symbol and form, are sourced in the breath and contact of living bodies. It offers an alternative figure bridging otherwise-isolated lived experiences and dimensions. Arguments against dualism abound, and this project is only intended to encourage practitioners across the disciplinary and pedagogical spectrum of work today to take themselves to task to think a return to sense, to the primacy of perception. It is not intended as a call to abandon the crucial and formative aspects of their projects but a return to the generative sources of their work. It is a return as a retrieval, and moving beyond, as a projection into a radical historical reimagining of existing practices, knowledges, and relations.

A new materialism requires that the meaning of theory needs to be revitalized. Revisited as the site for a primary reflective gaze and contemplation, classical *theoria* must now be saturated in the meldings of perceptual contact, affect, and the formulation of conceptual patterns. Breaking free from guidance by the ancient star of an abstract logos, detached from the workings of the image, as John Sallis argues, means also no longer teaching and learning notions of qualities as subsumed semblances of some essence or abstract algorithm or of absolute origins (even in mathematics) for all material and sensible things. For Sallis (2000) this requires a turn

> beyond the sensibly present to a coming profusion of sense, a profusion to come that indeed already comes, before its time, to fill out with all that imagination envisions the sparse traces actually seen. Proceeding from the sensible, it is a turning, not to the intelligible, but to a profusion to come, which already comes by force of the imagination. It is a turning within the sensible, a doubling of and within the sensible. It is a turning of imagination in an exorbitant sense allied in advance to the turning of philosophy at the limit (p. 73).

As an expression of sensibility, theoretical pursuits conjoin and integrate the formative powers of a radical social imagination. An elemental material curiosity, like the critical pedagogies inspired by Freirean culture circles, seeks to realize our affective and cognitive, perceptual and conceptual potentials. Theoretical curiosity is a name for the occasion of their alliances and phenomenal encounters; it is the gesture of the body in thought, of thinking's bodies in the world as it wakes to the problematics and puzzles of our habitation. We are placed in question as we inquire. As with ancient doctors, we are engaged in knowing our relation to and, in nature and the natural, of an always-given physis, in its writing into culture and mind. Like Socrates, the problems we must confront are the questions of the place of the human.

A premise of this project is that curiosity is an act of dislocating the familiar. A critical curiosity is always an act of suspicion or discomfort with the existing formulation of understanding and rules of engagement. The dislocations that motivate, guide, and are presented to research are recognitions of how strange the familiar can come to appear, whether suddenly, in initial vertigo, or a gradual methodological worrying of time-worn concepts, beliefs, or systems. The theoretical forays of Galileo, Newton, and Heisenberg in physical observation and theory come to mind. An often-volatile, though subtle and fluid force of thought, an attentive passion drives thinking across uncharted territories.

One instance of this can be illustrated in the conceptual production espoused in the work of Gilles Deleuze. Deleuze's theoretical and political explorations were intended as experimental productions of new linkages between thinkers and their traditions. Rejecting all humanist and representational historical accounts of philosophical thought, Deleuze's own heterodox projects redeploy an underground, alternative history that provocatively makes for unexpected connections between such diverse figures as Lucretius, Spinoza, Hume, Nietzsche, and Bergson. As an adventure of thought, his investigations are one multifaceted expression of critical intelligence reformulating perennial questions in new contexts and problematics. New objects of knowledge arrive, function, and come to be challenged in thought's irresistibly productive momentum.

A new materialism is a conceptual combat against institutional and cultural inertia. It is a constituted, historical stupidity that critical inquiry must always ardently oppose, engaging thinkers in an agon that, as John Rajchman (2000) emphasizes,

> starts not in some naturally given desire to know, or in accordance with a "natural light", but rather with the encounter with something that doesn't fit in habitual ways of seeing and thinking, that "shakes up" thinking and puts up something new to be thought (p. 44).

We hear again here the force of the anomalous, as in Feynmann's reflection on the initial "kick" of the new. This is present in the lived experience and experiment of theoretical forays in their quite contingent and irregular ventures outside of the given, the doxa, conceptual armor, laws and expectations of disciplined knowledge. The intent of curious seekers, harbingers of future critical communities, is, for Rajchman, to "make new forces visible, formulating the problems they pose, and inciting a kind of experimental activity of thinking around them" (p. 49).

I wish to point out the pertinence of some of the Deleuze's premises with regard to a renovated appreciation and practice of theoretical curiosity. Aspects of his thinking resonate with creating new figures for a non-representational appreciation of the productivity of theory and innovation. Though his work is largely averse to the received humanist notions of embodiment, and adamantly opposed to the early phenomenological interest in the constitution of subjectivity or "consciousness," many of his provocative tenets assist this refiguring theory and inquiry. I think it is useful to consider with him the vital signs of thinking about thought itself, both discursive and non-discursive, and whether mediated by signs or images, in terms of kinetic fields of energy and force. They are the quanta of cultural transmissions setting the contingent historical conditions of possibility for knowledge. They are the material, transactional relations of theory construction. This is a dynamic relation inherent to the materiality of thought. It is a kinesis that brings potentiality to actualization. As elemental forces within nature and the human, it is an *ontological productivity prior to or anterior to any epistemology*. It is the pulsing currents of thought. Theoretical and conceptual growth can only occur in the flows, turns, and circuits of thought and becoming. Images and signs mediate an ontological plenitude and excess.

In their collaborative inquiry into the nature of thought Deleuze and Guattari (1994) take on Spinoza's intimation of an immanent relation between a purely autonomous conceptual realm and its expression in the unlimited realm of infinity. They affirm that concepts "extend to infinity, and being created are never created from nothing" (p. 19). Aesthetic and philosophical inquiries conduct the vital work of forming new images of thought. Otherwise, a non-thinking repetition will occur through "functionaries" who repeatedly reproduce existent conceptual models, apparatus, and schema. In his dramatic restaging of the epistemological theatre of traditional philosophical practice, Deleuze and Guattari present a "*pedagogy of the concept*," an image of thought that would constitute a "new substance of being" (p. 51). In his own recomposition of Spinoza's precepts regarding immanence, substance, and modes, Deleuze (1988b, 1990) shifts the dominant twentieth-century traditions of philosophical orientation toward what he formulates as a non-representational becoming of difference. It is a unitary ontology of substance and mind that is not centered in individual subjects. It is a monism with a difference, one that always includes the copresence of the powers of thought and the power of being and becoming. Substance and modes, as thought through by Spinoza in the Ethics, are concepts whose consistency is bridge and continually rejoined within what he depicts as "planes of immanence."

The intensities of thinking to which curiosity provides kindling are inseparable from the consistency of "assemblages" or apparatuses of power. In a post-Kuhnian perspective, these assemblages or temporal forms are what Deleuze and Guattari (1988) identify as the "circuits, conjunctions, levels and thresholds, passages and distributions of intensity, and territories and deterritorializations" (p. 160). They are organized as moving conceptual modules, along vectors or "diagrams" formed in their trajectories and specific speeds along their "planes of immanence." A plane is not a concept "that is or can be thought but rather the image thought gives itself of what it means to think, to make use of thought, to find one's bearings in thought" (1994, p. 37). It is a description of thought's unceasing creation is intentionally transcendental. In this it would seem to be completely at odds with the phenomenological commitment to embodied practice. Deleuze and Guattari distinguish the transcendental nature of the planes where concepts proliferate from the transcendent. The transcendent is the dimension in which thought and concepts are in continual states of flux. Unlike the timeless transcendental, it suggests a layer, or in Merleau-Ponty's more limpid terms, a fold within our being. For my purposes his distinction is important in laying claim to the immanence of the conceptual "planes" to lived theoretical practice. These planes affirm an infinite fecundity for thought unbound from the clutches of humanist representation. Concepts and their conceptual "personae" mutate prodigally, reanimating disciplinary traditions, forms of pedagogy and the practices of knowledge in daily life. They make encounter continually with what has become the "problem" of life in modernity.

Thought oscillates, engendered within rhythms of time. It is a qualitative and quantitative, difference is its consistency. Curious thought and theory, in its meanderings or concentrations alike, conjoins through mediations, convections, and interceptions. Deleuze's Nietzschean-inspired antihumanist physics presents a thinking of all mediations, including the technological, as transversals, becomings, enfolding in textures of the known with the unknown. Key frames of reference can be reinvented in a new topology for thought. It can at the same time be formed at physical crossing points as in McNeill's (2005) gesture-image "growth points" to combine the energies of physiochemical, neurological, genetic, and ecological milieu (p. 82). Replacing the picture models of the world that Heidegger critiqued, entails rethinking the poles of representational mappings of the psychic "interior" with an exterior outside. Charged valences of intensity are constitutive of difference. They what may actually operate in what Deleuze portrays an infinite interplay between virtuality and materiality.

As an elemental aspect of our participation in the world, theory making forms essential aspects of our full engagement in it. It is one starting point for the continual interrogation of the nature of human existence. This is what Merleau-Ponty calls for when he asks

> If we are ourselves in question in the very unfolding of our life, it is not because a central non-being threatens to revoke our consent to being at each instant; it is because we ourselves are one sole continued question, a perpetual enterprise of taking our bearings on the constellations of the world, and of taking the bearings of the things on our dimensions. The very questions of curiosity or those of science are interiorly animated by the fundamental interrogation which appears naked in philosophy (1997, p. 103).

It is a "surpassing that does not leave its field of origin" (p. 153). It is the generativity of being, or in Deleuzian terms, the productivity of being that renders the visible as a *presentation*, a coming to presence in the realm of the sensible. Concepts, percepts, theory germination, and the "objective relations, the acquired ideas" like the scientific explorations of a Lavoiser or Ampere, or of the prototypical mathematics of Leibniz, are never independent of corporeal, relational experience. This relation allows the inventor of the calculus to "go straight to entities no one has yet seen, make the operative language and algorithm make use of a second visibility and make ideas the other side of" (p. 153).

Deleuze and Guattari, shaping their materialist pedagogy, claim that all philosophical pursuits, including my theoretical or elemental curiosity, participate in the production of these continually constituted planes of immanence. In attempting to think the "unthought" he provides notions of how this "absolute ground" of philosophy is a presence in the continual reinvention of concepts, ideas and systems of knowledge and identity. They will claim, enigmatically, that "the image thought gives itself of what it means to think. To make use of thought, to find one's bearings in thought" (1994, p. 37). All philosophical systems are mutable, moving cores centers for the generation of the energies of sustained inquiry along specific planes into which concepts assemble their contingent, but essential, "intensive" features. From Anaximander to the present, any philosophical work is

> at once concept creation and instituting the plane. The concept is the beginning of philosophy, but the plane is the instituting. The plane is clearly not a program, design, end, or means: it is a plane of immanence that constitutes the absolute ground of philosophy, its earth or deterritorialization, the foundation on which it creates its concepts. Both the creation of concepts and its instituting of the plane are required, like two wings or fins (p. 41).

This perspective resolves sensibility or embodiment with the intelligible and conceptual. Theirs is a reciprocity and continuum of differences in continual emergence. It does not mean that the empirical, sensory, or aesthetic body operates at all times, governing "from below" the generation of thought. From the poetic sensibility of Merleau-Ponty's phenomenology it is a layer or strand of visibility, a "second life," an invisibility or horizon of possibility not immediately apparent to unreflected experience. He describes a two-leaved body, one touching, sensing, and seeing things, and the other a thing among other things. It is a clasped contact, like that of one hand holding the other, between a sensible and a sentient corporeality. Ideas, mind, and thought are in this sense of sensibility a "sublimation of the flesh" (1997, p. 145). In his reading of the visible world, our body is composed of the same flesh of a world of depths, of a latency and potential:

> the visible is. . .a quality pregnant with a texture, the surface of a depth, a cross section upon a massive being, a grain or corpuscle borne by a wave of Being (1997, p. 136).

The visible of our perception is always enveloped by the invisible. The potential dimensions of experience, perception, and knowledge are "pregnant" with an infinity of the invisible. It is an intercorporeal relation, one that provides for the counterintuitive notion of an inherent correspondence between the world's exterior and our bodies inside. As a surface of an inexhaustible depth and abundance, the body as flesh of the world is also the presence of the absence into which the horizons of thought come to presence. They present active, volatile sense images, perceptual filaments and tangling strands; as this second visibility, it is a process of thought's becomings, one in which it is always "brought to appear, we say and not brought to birth" (p. 145).

As the urge to adventure, to challenge known physical, intellectual, and theoretical frontiers, the persistent spirit of inquiry evinces cultural anxieties and ambivalences regarding the limits and durability of human understanding. The elemental spirit of inquiry is a marshalling wave and point of convergence for an excess of cultural desire. Daedalus-like figures abound in oral and written cultural legacies as those who would venture and test the limits of human finitude. Socrates, who claims his ignorance in the *Meno*, is an exemplary persona as a seeker in the face of this encounter. Socrates professes ignorance in acknowledging his mortal limits.

For M. Dillon (1998), in his "endless search to know, he is exemplifying *aperte*, the ideal way for man to respond to his finitude" (p. 3). As a restless seeking for knowledge "sustained throughout life" the human condition and its desire for knowledge will always be unfinalized. Dillon reads this dialogue as a strong claim for recognition of our finitude within the philosophically disciplined life dedicated to human affairs, rather than cosmic or metaphysical questions. Socrates responds to Meno's claim that knowledge, like becoming, will always veil an inaccessible realm of pure being. His response is both skeptical and affirmative. For Dillon it amounts to a recognition that

> human reality, the only reality to which we have access as finite beings, is the correlate of human knowledge, what *is* for us is such as to admit of continuous unfolding, continual discovery (p. 3).

Blanchot (1993) poses the questioning of our experience as an ontological search. It is the phenomenological journey generated by the "question of origin." It is a search for the foundation, ground or contingent "truth" veiled within the phenomena of the visible world, and the relentless movement of images in the totally mediated world of the spectacle. For Blanchot we interrogate ourselves about everything, in order to sustain and advance the passion of the question, but all questions directed toward the one question alone, the central question, the question of the whole (p. 11).

It is a libidinal quest, a question posed by desiring knowledge. Recalling Diotima in Plato's *Symposium*, it is an essential loop within the erotic nature of knowledge. The origins of the question of the whole is also the eternal cycle of return, in which knowledge lacks an origin outside of its own rootless desire. Desire surges within and between humans and the material phenomena of experience. The loop is the lack within an always partial knowledge, order, or representation. It is an ontological movement that gestates any historical epistemology and its imprint on our ways of knowing within and outside intellectual and material communities of practice. For Bernard Steigler (2009), the question of the origin as the question of the whole is also always marked as the "question posed by desire which does not cease to question and the question that poses itself from the moment when one questions desire, when one asks of desire what it is" (p. 8).

Merleau-Ponty called the link between desire, affect, and the intellectual or discursive aspects of human expression "representative sensations." By this he was indicating that there is an "affection" or presence in and to the world inherent to the body and through it. These representational sensations move with the sense of a language porous to our membranes, through the "world being flesh." Language is not exterior, but presents an "expressive teleology" within our "lifeworlds" layers and the tissues of corporeality. Merleau-Ponty stresses the place of affects in the expression of concepts and representations. He claimed that the "structures of affectivity are constitutive with the same right as the others, for the simple reason that they are already the structures of knowledges, being those of language" (1997, p. 239).

For Barabas (2004), there is no meaning, statement, deduction, or principle that is not "incarnated, and, in this measure, affection. There is no affection, particularly in the relation of desire that is not already the advent of sense, an attempt to carry the world's opacity to transparency" (p. 278). Visibility and desire are always commingled here as in the question for the whole, of the origin or the totality behind or within the phenomenal spectacle of modernity. In Barabas's rethinking of what Merleau-Ponty sought after with his tentative premises regarding the flesh, desire is the "universal power of incorporation, flesh open to flesh." The visible, sensible world is only that which has come to emerge out of the infinity of horizons, perspectives, and virtualities of becoming. He claims that at the "heart of the visible sense remains invisible" and that "everything that can present itself is given only as absence" (p. 278). His notion is not dualistic, marking a divide between the phenomenal and noumenal worlds. Phenomenology always attended to intentional acts and to the questioning how things come to appear for and to us. The presencing of things and relations precedes all representations and discursive signs. The coming to presence is always sparked by the desire to see, rather than having seen, to use Heidegger's phrase. Any presencing is an act of desire. It is the pursuit and opening up, carving out, or working ideas from form, as in Leonardo, of a knowledge's pursuits.

In the modernist European tradition, from the rise of empirical science and the Enlightenment's positing and pursuit of universal values, the realm of the given, the everyday, and the sensible remains suspect. Upholding, verifying, and subject to experimentation, it is often a respectful suspicion that informs the attempts, innovations, and research programs that conserve disciplinary knowledges while accumulating new empirical "evidence." In the Kuhnian model, these processes authorize departures from standard models, tentative methodical shifts toward uncertain programs and paradigms. The limits of what is knowable are not surpassable while the same models, compasses, maps, or grids of intelligibility are retained. Models are useful only so long as they remain uncertain. Social theory, and for that matter, social life, as discussed in my last chapter, is the basic context for this contest of existing representations and models.

The relation between the sensible and the intelligible is a fold, a knot, or reversible circuit. For Merleau-Ponty, any claim of a surpassing into the conceptual or "noetic" must always be recovered by a fruitful return to its origins in the lived textures of experience. There are certainly qualitative differences in the quality of experience expressed in linguistic signs, symbols, propositions, and concepts from perceptual sense. Yet there always remains a continuity between them. From the phenomenological perspective it is a distinction and difference within a unity. For Merleau-Ponty this is the distinction between forms of visibility evident in all of humanity's vaunted cultural inventions and interventions. Though this distinction can and must be recognized

> it is too soon now to clarify this type of surpassing that does not leave its field of origin. Let us only say that the pure ideality is itself not without flesh nor freed from horizon structures: it lives of them, though they be another flesh and other horizons. It is as though the visibility that animates the sensible world were to emigrate, not outside of everybody, but into another less heavy, more transparent body, as though it were to change flesh, abandoning the flesh of the body for that of language (1997, p. 153).

Announcements of a passing beyond of the limits of human knowledge and relation have been declared in all eras. They affirm visions of transcending borders and frontiers. Knowledges will always be partial, expressive of the contradictory interests motivating their historically situated pursuit. Knowledge is surpassable not through transcending the boundaries of a historical apparatus, but by reconfigurations of historical models of producing material relations.

Intentionally purposeless in relation to epistemological truth claims, projects of sustained intellectual inquiry suspend totality, utilitarian aims, and wills to system. The kinds of inquiry I wish to champion are instances of a willful and motivated deflection from solely utilitarian recuperation by established modes of knowing. Like their medieval counterparts in the vilified affect of *acedia,* practitioners of an elemental curiosity divert purpose, goal, and teleology. Straying or lingering in empirical uncertainty, these idle researchers are students of chance and contingency and, like the Situationists, they await and engender unanticipated convergences and complexes of forces. Skeptical toward universals within the natural and human worlds of experience, they are decoders of static and noise, pulsations of heat and movement within any system. In basic natural science research, inquirers like Heisenberg, Gell-Mann, Sagan, and Feynmann, for instance, exemplify a sustained care and attention for the disparate. Their researches reckon with the limits of existing understandings of the intricacies of a universe of complex coemergence, and integrated subsystems that disturb foundational unifying theories.

It has been a primary contention of this project that intellectual processes of inquiry are forms of distinctively accented, historically determined and interested curious activity. The questioning definitive of any sustained critical inquiry is a potentiality and productivity coming to appear in interventions. Their undetermined reserve can unground fundamental principles and orthodoxies. The objects of knowledge are continually produced and reshaped by human material and intellectual labor. In Gilles Deleuze's response to the philosophical conversation regarding the question, "what is writing?", he offered this observation:

> the question silences all empirical responses which purport to suppress it... whence the power of the question to put into play the questioner as much as that which is questioned, and to put itself in question (1994, p. 195).

As a material intensity, intellectual activity and thought's abundant researches are contingently summoned through differences and differentials. Curiosity constellates, continually shifting and stabilizing within representational and discursive closure. Its intensities, active and reactive in a Nietzschean sense, leave traces that can be mapped into what Peirce called "habits of embodiment." These qualitative and quantitative differences of empirical sense and experience determine the existing conditions and kinds of experiences subject to persistent revision. They also pose new or alternate modes of questioning. In all discursive fields of experience, thought and language are always mediated, plied, and playing within overlapping, competing and often rival interests. Experiences of thought are material, not in the misunderstood sense that would make them parallel physical properties, but as constituent elements that shape and operate between dialogic partners, in some actual or potential community of inquiry.

Thought oscillates, engendered within rhythms of time. Qualitative and quantitative in their differences, the meanderings and concentrations of sustained inquiry conjoin through mediations, convections, and interceptions. Mediations are generated in transversals, becomings, enfolding in textures of the known with the unknown, same and different, internal and external, inside and outside as they are comprised in the conventions of boundary marking in disciplines. Changes and the charged valences of intensity are constitutive of difference. The singularities of inquiry are linked by the will to disclose, discover, unravel, or dislodge the familiar. Traveling light in the modern *episteme*, curious mediations pursue trade winds and venture off course, making contacts, links, and connections that are always somewhat unexpected and unusual, bringing surprises or news that might not necessarily be readily useful. Curiosity's ship navigates by stars even under cover of cloudy skies.

As a chimerical, yet enduring form of desire, curiosity finds itself working in unsuspected places, usurping the orderly rationality that it tracks and shadows. It is an instance, always particular, describing an arc of inquiry, an empirical design taking shape. It is the activity of the sensible, of the aesthetic makings of human production, a humming loom for the textures of experience. It is empirical, speculative, and theoretical, never reducible to a measurable cognitive status. Turning on its own contexts and pretexts, it moves, auto-formative. The dynamics of this passion of the mind are fueled by the impressions lived experiences impart. Curiosity does not explain itself. It appears as if without history or paternity, a taproot, a lady's slipper, or rhizome, making its appearances without visible or audible origin.

Chapter 6
Thinking Life

The man of the future will be filled with animals
—Rimbaud

In this and the next chapter I am concerned with giving special attention to the theoretical curiosities and social energies apparent in the technosciences. In this chapter I will profile the cluster of research programs that emerged with the race to decode the human genome. I offer a brief historical context for the "genomic" model's originating sources and research purposes. I give priority to the technosciences in general, and to the bioinformatic, informatics, and astrobiological research programs (in the following chapter) as specific historical manifestations of *technics*, or any of the unlimited forms of all human practice utilizing tool use, instruments, and devices augmenting our biological and intellectual capacities. Raising historical and anthropological perspectives to ground this inquiry allows me to draw out a few of the multiple and, ultimately, indeterminable facets and effects of these new interventions. Genomics, as a much-vaunted biotechnological practice, is a portmanteau term for a diverse cluster of theoretical projects that bring questions of embodied experience and the place of scientific curiosity to the surface. I wish to encourage a historically nuanced consideration of the often-hyperbolic promises and processes actualizing in the technologies in their potential capacity to implement specific, though often diffusely defined, social interests. The hybrids of technoscience, as represented by the fields of genomics, proteomics, and nanotechnology, demonstrate their ability to transform existing forms of cultural recognition, memory, and embodiment in *everyday life*.

The nature of everyday life is presided over by modernity's technical expansions. Research and theoretical pursuits have inherited a visualizing epistemology. It remains dominant, though contested, reigning in many intellectual and academic disciplines, especially the nearly ubiquitous Anglo-American family variants of philosophy of mind and language, including emerging fields like those represented by students of "extended mind" and neurocognition. An epistemological will to power has hollowed out a pedagogical vacuum without attending to the material world of our senses and the perceptual grounds of knowledge and theoretical growth.

I believe that we are at a crossing point between eras and systems of theoretical practice. We are heirs to an era in which representational and visualist models and

M. Zuss, *The Practice of Theoretical Curiosity*, Explorations of Educational Purpose 20, DOI 10.1007/978-94-007-2117-3_6, © Springer Science+Business Media B.V. 2012

methods have held supremacy. Securing the epistemological grounds for knowledge has shared sovereignty with representational models, co-emerging alike in mathematical, rationalist, idealist, and empiricist guises. Representational "maps" of knowledge and the frames of its objectivity are projected from the scanning eyes of its spectating researchers. They are both fortified by an unprecedented magnitude of technological rationality and instrumentality. The conditions of their possibility, deeply engrained in paths of inquiry since the seventeenth century at least, have been secured by the economic and political structuring of entrepreneurial, and then industrial capitalism, in its extraction of its epistemic objects, axioms, and laws from lived phenomenal sense. Among the many effects of this tradition has been to eclipse the value of the Orphic curiosity of the aesthetic, the epistemic over the perceptual and material labor. Manual work, whether artisanal, in craft or manufacture, still lacks the cultural cachet attached to intellectual and knowledge work. Technological forces, carrying these dualist legacies, engineer much of the current direction for theoretical investigation, which any genuine critical practice and thought must resist.

The focus on the technosciences is intended to further attention to a reflexive stance toward dominant modes of contemporary inquiry. I have taken as my task the questions Paul Rabinow (1996) raised in his ethnographic participation with geneticists, microbiologists, biochemists, and ethics committees considering the purposes, directions and limits to research. Rabinow called for a reflexive practice that appeared absent in much of the molecular biological work. There was no internal principle for the consideration of the question of limits. Pointing out that reflexivity can merely be confined to the making of normative scales and methods to operationalize terms, Rabinow regards it as an "an experiential and experimental 'problem'." Posed as a question of the "usefulness" of research

> this stance entails being curious about scientific curiosity, and curious about one's own curiosity. It leads one to thematize the form of life that surrounds, sustains, and undermines curiosity. Thus, even when claims are made to have discovered 'the curiosity gene,' the question of what kind of society has posed such questions to itself, why it has sought to produce this type of knowledge, will remain open (pp. 170–171).

The ancient doctors, despite their different idioms and doctrines, shared an abiding interest in the manifestations of natural processes within both forms of life and the inanimate. Two key terms deriving from the classical context of Aristotle's investigations of the genesis of living and natural forms return. Aristotle's formal premises and strong reinterpretations of his predecessors generated a rich tradition, revived in the late medieval period, of natural philosophy They remain pertinent to the new technologies and forms of theoretical practice evident today in biotechnologies, in its varied and cross-disciplinary fields of research. *Physis* concerns the self-generation of forms of nature, of a primordial and autonomous genesis of things, and of forms of life, in particular. *Techne* or technics signifies the historically available and useful instruments, tools, and methods that allow for human intervention into nature. Technics is contingently manifested in the apparently unlimited potentiality of contemporary technological forces of production. It represents human intervention and the realm of practice in the arts and sciences. Techne usually refers to the fashioning of materials. These can be marble, wood, cloth, whether in manufacture and making of things starting with a design, model, or image. It consists of distinctively *human* work and crafting of materials.

Since human practices are set into place through the use of tools and instruments, the technical here refers to an "artificial" means of working in bringing an object, idea, or material to actual fruition. It includes, in its original sense, theoretical activity. As an expression of a primary act of bringing things into appearance by human craft, they are creative acts of making or *poesis*. The technical in the classical context refers then simultaneously to artisanal and intellectual forms of making things. With regard to theoretical or intellectual inquiries it indicates the capacity of bringing the potentiality of thought into actuality, the making of new concepts and unities within difference. *Techne* is the realm of intentional human interest and interventions into and with *physis* or natural and organic processes, including altering human embodiment and perception. *Techne* is also a field of representation, of design in the fashioning and crafting of objects of inquiry and consumption.

Technics, as the historically specific extensions of an epistemological will to certainty in knowledge and the deployment of power, has come to define the structure and processes of contemporary social relations. The concept of naturalness is under trial, exposed as a carefully calibrated, historical effect of human intervention. Nature is armed.

Genetic Capital

Technical forces are productive and produced, working through human and nonhuman relations. They find actuality through the coordination of abstracting of the practices, techniques, and instruments of the biosciences. The place of technology in contemporary culture invites vigorous scrutiny of its specific functions and historical articulations within and between institutions, in ensembles of social practices and forms of constituted experience and subjectivity. Material culture is complexly configured enough to engage several, sometimes rival, modes of technology and ways of knowing the world while retaining social cohesion.

"Posthuman" and "transhuman" specters haunt contemporary social discourses. Discussed here and in the following chapter, they serve multiple agendas and interests. The posthuman, if it is arriving, reprises cultural processes by which the colonial and Republican citizen-subject was given coherent form. Incising new boundaries and extensions of embodiment, I propose that the advent of new biopolitical technics presents one means by which the grain of daily life might be recoded and administered. Microbiology, genetics, and computational cybernetics share close historical research and institutional alliances. They are technologies of the self, forays along certain currents of theoretical practice in the natural sciences. As active expressions of a particular theoretical curiosity they challenge key modernist values and concepts of the natural, the human, and the epistemic status of the subject. The biosciences include their own hybrid investigations, forming their own evolving taxonomy of practices. The biosciences are an ensemble of practices that have operated under a panoply of rubrics, including the cross-disciplinary work represented by genomic, proteomics, bioinformatics, complex systems biology, and others. The techniques, instruments of these "life sciences" have adopted computational and complexity models which may become paradigmatic with the gradual rise of systems biology research and funding. Complex systems biology has largely succeeded basic genomic research programs like the much heralded and apparently hyperbolic historical claims made during the heyday of the National Institutes of Health–funded Human Genome Project. Few, if any, direct medical treatments and therapies have resulted from the multibillion project that Francis Collins (2010) claimed was the "most significant thing humankind has tried to do scientifically" (p. 65).

The new technics appear as rationalized instruments for the privatization and management of the range and choices of "natural" embodiment. Working in alliance with private health care "options," they are capable of setting the terms and available forms for regulating reproduction, health, illness, and the image of the human. The dominance of fMRI diagnosis is just one illustration. As contemporary forms of technics, the genomic, and bioinformatic paradigms are specific assemblages of practices, methods knowledges, and apparatus representative of what Paul Rabinow calls a "biosociality," affecting the textures and qualities on the human fabrication of its own future (p. 87).

A general cultural anxiety has been stirring since the 1990s in debates over the potentials of the new technics. Self-appointed defenders of an illiberal democracy like Francis Fukuyama and Dinesh D'Souza appeared on the defensive in the face of a technology that exceeds the deregulated "posthistory" of an unfettered free market. Fukuyama (2002) stood in opposition to those proselytizers of a liberating technophilic posthumanity, like Donna Haraway. Despite his own influential declaration of the "end of history" after the fall of the Berlin Wall, he piously affirmed the perduring value and existence of a core human nature, asserting that

> provided a stable continuity to our experience as a species. It is, conjointly, with religion, what defines our most basic values. Human nature shapes and constrains the possible kinds of political regimes, so a technology powerful enough to reshape what we are will have possibly malign consequences for liberal democracy and the nature of politics itself (p. 6).

The new genetic and digital technologies are specific and determinate forms of contemporary biopower. They represent new codifications of disciplinary practices and forms of life management and the erasure of the public health choices. Considering the thought of a purportedly "posthuman" subjectivity in phenomenological terms opens up the effects of the diverse encounters between the human and the informatic, the genome and the cybernetic management of subjects. In the following sections I survey aspects of the genomic, bioscience, and artificial life paradigms as actualizations of technics. They present varied ethical, practical, and theoretical potentialities. Their public rhetoric usually announces a surpassing and redrawing of the limits and relations between the human and the non-human, the artificial and the natural worlds.

Whether through somatic cell line, or more auspiciously, through germ-line therapies now capable of altering the transmission of genetic makeup of offspring, the limits of the human are transforming. The concept of the human as a discrete, singular biological species is itself mutating. Genetic configurations of the possible forms of embodiment are prevalent; they are evident in developments extending the muscular and nervous systems, in prostheses and hybrids of bacteria and host, hominid and machine, cell and digital chip. In addition to cell lines and genetic material, transgenic plants and animals are being developed and introduced into the ecosystem where their interactions cannot be adequately anticipated. Hybrids of hybrids, of the genetic legacy, monoclonal, recombinant DNA cloning, and somatic and germ-line technologies are knowledges that reorient the relations and forms of *zoe*, nature, and its social and individual embodiments. Acting through a microphysics and micropolitics of bodies, a biosocial episteme is inaugurated in the activating theoretical forays of this techne, their potential realized, as the power of the molecular to act upon other molecular bodies.

Prevailing discourses of genetic engineering are hyperbolic in their claims for human enhancement and disease control. Patenting genes that might promise drug therapies makes the molecular structure of what we have known as human the private property of biomedical and pharmaceutical multinationals. This can readily yield to conditions of control as extensions of existing medical and technological management of human populations—setting the conditions for who thrives and determining the countenance, stature, and image of the "posthuman." Like their parallel "public" aspect in the increasing practice of embedding RFID chips in animals, children, and sex offenders, biopower's current technics represent very augmented visualizing magnitudes of surveillance. Centers formerly housing nuclear weapons research and development, including the Lawrence Livermore, Los Alamos, and Oak Ridge laboratories, were retooled, using their unparalleled IBM supercomputing prowess to track and record the genome. They have functioned as technological points of production of new knowledges and powers in a micropolitics of genetic agents, alleles, gene markers, bacterial hosts, vectors, and mutation flows.

Genomic potentials participate in a generalized contemporary vertigo of identity, doubly confounding modern biological and sociocultural forms of human recognition and regulation. Beginning with the race to decode human and animal DNA, their interventions have disturbed formerly sacrosanct divisions between the human and the animal, as well as between the apparently fictive borders between the animate and the inanimate. Fabrication of genetic material, in transgenic animal crops, embryonic stem cells, polymerase chain reaction synthesis of DNA, and the increasing variety of forms of reproductive technologies, including cloning in the animal world, all call attention to an alignment of the technics for transforming the "nature" and relations within and between bodies and populations. No longer regulated by archetypes of race, or blood purity, the new physiognomies of the human are being transfigured in the micromanagement of cells, genes, and the vitalized "information," including the DNA and retinal "profiling" being introduced by police agencies.

The genomic paradigm prepared the ground for a technical theoretical curiosity in which new differentiations and potentially unlimited (and unregulated) integrations became possible. Germ-line therapies in particular have presented challenges to the shape of future bodies and their constitution. In September 1998, the Center for Responsible Genetics, which monitored genomic research initiatives, announced an "action alert" in response to W. French Anderson's proposed projects, as reported to *Business Week,* at the University of Southern California School of Medicine in Los Angeles. Anderson's team attempted to cure both a rarely occurring immune disorder and a form of anemia, both caused by genetic defects, by the insertion of new genes into the fetuses of animals. Anderson's proposal extended to treatment of humans. Alterations to the somatic cells are capable of being carried over to the reproductive, germ-line cells. Traditional and modernist representational tableau of gender, race, and class are being challenged by forms of genomic biopower, as the emerging, if contested, practice of germ-line therapies makes evident.

Lily Kay (1993) places the development of a "molecular vision of life" in the institutionalized research agendas established through the Rockefeller Foundation. Kay examines the emergent molecular and genetic paradigm through the concentration of capital and resources at Caltech in particular in the 1920s and 1930s. Redirecting biological research from the sullied aftereffects of explicitly eugenics orientated programs and the blemish of the *Buck vs. Bell* decision allowing for forced sterilizations, the newly emboldened "Sciences of Man" nonetheless set out to create a "social physics," intended to produce a coherent body of applied research concerned with the control of contemporary social relations. Kay claims that the Rockefeller Foundation's "officers and their scientific advisers sought to develop a mechanistic biology as the central element of a new science of man whose goal was social engineering." The popularity of programs for social or human engineering arose in the early years of the twentieth century. She documents how the term "human engineering" resonated with the Progressive Era's faith in technological progress and a belief, at least among the scientific and intellectual elite, in its ability to transform human nature.

> Midway during the 1910–1920 decade the term "human engineering" came to denote the application of scientific principles and technical methods to social and educational process associated with the maintenance of social order: stable families, work groups, and rational management of changing sexual and racial relations (p. 17).

The new molecular vision was explicitly instrumental. Kay stresses the fact that the Laura Spelman Rockefeller Memorial Foundation's philanthropies made clear that the funding was reserved not for

> the promotion of scientific research as an end in itself; its motive was not *sheer curiosity* as to how various human and social phenomena came to be and are; the interest in science was an interest in one means to an end (p. 48).

For Kay, the Rockefeller-generated science agenda was set into motion as the "surest foundation for a fundamental understanding of the human soma and psyche—and ultimately as the path to rational social control" (p. 50). Its pseudoscientific project was nothing less than what she identifies as a "taming of the savage."

The intellectual curiosity driving the organization and development of research agenda has been explicitly Promethean and instrumental. It has seized the means and ends to exploit, harness, and control the definition of "life sciences." In constraining the limits of the human and its rational organization within a technoscientific reduction of the meaning of natural causation, the molecular vision and its genomic descendants continue the projects of human and animal engineering. Codifying the natural and the animate in determinate regulatory patterns accelerates the means and ends to any interests that seek to restrain, expand, or blur the boundaries of the "human," the organic, and the animate. An enduring human engineering agenda, one that is heavily freighted with the specter of former eugenic projects, pervades the development and social reception of these fields. Genetically modified foods (GMFs) and high-tech agribusiness patents consistently meet strong and persistent local resistance, including raids on cultivating fields, as in the dramatic initial negative reception of Monsanto's Roundup-Ready Terminator gene in India and elsewhere.

Under Max Mason's directorship, the Foundation shored up projects that sponsored the "salients" of concentrated, interdisciplinary research programs. These have catalyzed capital investments and interests perfecting the means for regulating the reproduction, behavior, and conditions of labor. Mason waxed poetic on his "salients," constituting research efforts

> directed to the general problem of human behavior, with the aim of control through understanding. The Social Sciences, for example, will concern themselves with the rationalization of social control; the Medical and Natural Sciences propose a closely coordinated study of the sciences which underlie personal understanding and personal control. Many procedures will be explicitly co-operative between divisions. The Medical and Natural Sciences will, through psychiatry and psychobiology, have a strong interest in the problems of mental disease (p. 46).

Among the questions raised by historians of the molecular and genetic research projects—Kay, Rabinow, and Keller, among others—is the situating the human within the molecular and genetic paradigms. For Kay it is an issue of thinking the "directionality and modality of seeking knowledge" (p. 16). This line of inquiry traces how the molecular vision and its successor in the genomic paradigm are extremely reductive. Their scalar models concentrate epistemic attention to identities at extraordinarily small units of activity. In the Rockefeller Foundation's archival statements this is emphatically endorsed in their concern for previously unknown levels of analysis made possible by the new centrifuges, electron microscopes, and spectroscopic tools being developed and implemented with the research programs. In particular, Kay points to molecular biology developing its own representational and optical "locus of life phenomena" between the region of 10-6/10-7 cm levels. As she points out, this "region was the main functional domain of the new biology" directing and determining the "form and content of research." Under Warren Weaver's stewardship, a kinship to physical models was ensconced allowing for analogies between cellular and subatomic organization. Research was conducted with attention aiming its scanners to the "ultimate littleness of things" (p. 49). Kay makes explicit one my key concerns when she wonders if

> question arises whether life is molecular or it is only the vision that is molecular. By studying macromolecules, do we study the salient attributes of life, or only a molecular representation of life, one of many possible representations of animate nature? (p. 16).

Technics and Culture

Curiosity as a spring of knowledge of the world and the extension of human material existence is an effective power of thought. Participating in the reinvention and extension of the conditions of daily life, its mutability works and worries the edges of existing technologies of representation and identity. It is a force of intellectual engagement and inquiry into the durability and value of any existing means of reproducing the cultures of daily life. Theoretical curiosity contributes to what Marx (1992) asserts in the *Economic and Philosophical Manuscripts*, writing about the

> fabrication, through human labor, both intellectual and physical, of the human sense, the humanity of the senses—all these come into being only through the existence of their objects, through humanized nature. The cultivation of the five senses is the work of all previous history (p. 352).

Technics represents the human cultivation of the body, the art, and disciplining of the senses. It is what Marx identifies with an emancipation from estrangement within capitalism's abstracting and extracting from generative social relations, in particular the reign of private property from which he is able to construe how the

> senses and attributes have become human, subjectively as well as objectively. The eye has become a human eye, just as its object has become a social, human object; made by man for man. The senses have therefore become theoreticians in their immediate praxis. They relate to the thing itself for its own sake, but the thing itself is an objective human relation to itself and to man, and vice-versa (pp. 352–353).

The sensuous and perceptual energies of the human are harnessed into productivity and utility. Social formations, cultural communities (including scientific cultures), and their reciprocal practices are appropriated; seemingly autonomous technical, automatic, and non-human means drive their prowess over the animate lifeworld, territorializing modes of production, distribution, and circulation of the material yield of human work. Human praxis and the desire for knowledge are rationalized; its potential pedagogical ventures are instrumentalized and codified in protocols, techniques, and instruments for their management.

Technics, as the forms and capacity for extending human sensory and intellectual abilities through tools, devices, and artifacts, has always provided a fulcrum and point of convergence for interest, fascination, and innovation. It refers to more than the technical instruments and artifacts cultural groups fashion in extending the hands, eyes, ears, as well as the torsion and strength of human bodies in their active production and reproduction. Since at least Lucy in the Olduvai Gorge more than a million years ago, technicity has conditioned human survival, passions, and possibilities. It manifests them, as they are exteriorized in a culture's available tools, apparatus, and instruments ready at hand.

Utilizing the paleontological investigations of Leroi-Gourhan (1993), I wish to trace, in a genealogical sense, the place of the genomic paradigm. It is my intent to question the claims for a surpassing of the human, as a mode of subjectivity, within a historical and anthropological consideration of technics.[1] Technology and culture have always coexisted, reciprocally enabling the pursuit of specifically human extensions of the species' knowledge and power over nature and itself. The genomic model and its apparent successor, complex systems biology, have posed an encounter with ethical and ontological grounds by forcing critical attention on the question of the limits of the species' discrete integrity and identity. Our biological embodiment, in its qualitative capacities and functions, is increasingly susceptible to mediations and mutations. Whether the productions of the new technologies are ameliorative or destructive will remain an open question subject to democratic participation in their deployments from the laboratories to the clinics, through the informed involvement of doctors, patient communities, and managers of health care.

As cultural expressions, technics shapes contemporary innovation. They extend, while confounding, foundational tenets of modernist subjectivity, embodiment and experience. They are distinctive in their power to produce novel artifacts, values, and forms of (dis-)embodiment. Carl Mitcham (1994) calls for a "history of ideas about technology—that is, the study of how different periods and individuals have conceived of and evaluated the human making activity, and how ideas have interacted with technologies of various sorts" (p. 116). He provides a starting point for such an inquiry, presenting differentiating models and the meanings of techne from antiquity and the Classical world.

From an anthropological perspective, technics are inherent to material culture and embodiment. They are forms of exteriorization of the body and thought. Whether in agricultural, and in increasingly transgenic products, implements or tools for hunting, carving or smelting metals, they are present in differentiated technological systems. They are the materially available means for particular cultures to organize, appropriate, and develop aspects of environmental niches and into which they are inseparably integrated.

Bernard Stiegler (1998) presents technics in originary terms. It is a formative matrix for social relations. Embodying temporality it reincorporates thinking's anticipation and a sense of futurity. For Stiegler and Richard Beardsworth, who both draw on the paleontological work of Leroi-Gourhan, technics are never exterior to cultural form, practice, or expression. In this pan-human perspective, they comprise the potentials afforded by the physical, intellectual, historical milieus from which prevailing, discursive practices develop. Thinking the history of the human and the possibility of its surpassing into the "posthuman" involves a perspective appreciative of differences in how cultures organize their activities in reproducing material and symbolic systems. For Stiegler, a complementarity is at work insofar as the technical is perennially "inventing the human, the human inventing the technical" (p. 137). The emphasis here is on the presence of technology as extension, prosthesis, and supplement. Leroi-Gourhan makes the bold claim that "the whole of our evolution has been oriented toward placing outside ourselves what in the rest of the animal world is achieved inside by species adaptation" (p. 235). This is to accent a modal, developmental perspective on technical evolution, or as Saul Ostrow (1997) phrases this in the context of digital communication technologies, our "becoming the object of technological developments that were once secreted within our body" (p. x). As an expression of the technical tendency to exteriorize memory, Leroi-Gourhan argues that

> our techniques, which have been an extension of our bodies since the first Australanthropian made the first chopper, have reenacted at dizzying speeds the events of millions of years of geological evolution until, today, we can already make an artificial nervous system and an electronic intelligence (p. 173).

Beardsworth considers technics as immanent to knowledge production. As a groundless ground, technics provides the conditions of possibility allowing for signification and representation. As in Derrida's (1976) depiction of the *gramme* and program, each material inscription, or bearing of material marks, is an actualization of the virtuality of an originary "arche-writing." It is the supplementary, exteriority of becoming and difference. Beardsworth writes:

> following the thesis of "arche-writing", technicity is the originary supplement to all forms of life, this supplementarity appears as such within the human species qua the technical specificity of the human. As a result, technicity can be reduced neither to the structure of a means nor to that of instrumentality, as the metaphysical tradition from Meno to Heidegger conceives it. It is the way in which life lives (p. 79).

At the same time, any generic typology should be replaced through critical work articulating the "differences between various forms of technicity (genetic, cellular, organic, inorganic, etc.)" (p. 79).

Temporality is a primary aspect of technological systems. For Stiegler, technicity is the matrix in which all experiences of temporality are given contour. In his reading it is a primary and structuring evolutionary framework for representation. It organizes and patterns material cultures' spatial and temporal relations. The future is invented in the present through technics in "structural couplings" that defer any originary exteriorization process. The modernist epistemic subject historically crystallized as a form of split being composed by a partition between experience and reflection, between interior and exterior domains of knowledge and relation. Stiegler's rich rendering of the anthropological record as trace, or Derridean *gramme,* entails appreciating the significance of the "question of the name of man" is the grammatological question of *difference*, the default or lack of origin from which any externalization could ever emerge. The gramme "structures all levels of the living and beyond, the pursuit of life by means other than life." (p. 84). He quotes Derrida (1976), whose work also drew from Leroi-Gourhan, asserting that the *gramme*, qua technics, structures all "genetic inscription up to the passage beyond alphabetic writing to the orders of the logos and of a certain homo sapiens" (p. 84). It is a reading that brings a complex interweaving to a whole tradition and culture of humanist enclosures of interiority and subjectivity.

Abstracted by codes and protocols, language is becoming the parlance for human surrogates to the technization of all social relations. Deracinated from physical presence and experiential contact that are not entirely mediated, language becomes the sounding of ghosts. Individuals, groups, community affiliations, "identity kits" are relocated to nowhere, as when a generic GPS voice directs the coordinates for how to move through our lives. Technization has become an almost autonomous power to affect bodies, relations, concepts, and historical relations. Dominique Janicaud (1994) is finely attuned to how language itself, structured through technological practice and forms, is becoming rootless. The productivity at work in the ascendant electronic techne "threatens the richness of embodied life and its gestures." The digital machine accelerates a detachment from the body, from any trace of the particulars of experienced place and time. As a "detached human gesture," electronic technology "becomes the object of everyday language, while at the same time at the level of the transformations of projects, mental attitudes, and vocabulary it is to a greater degree its driving subject" (p. 89).

Janicaud summons Merleau-Ponty's lyricism when he questions the viability of living language and of lived space, now standardized, and "manipulable at will" in a cultural confrontation exponentially maintained by electronic media and information-systems:

> A language rooted in the obscure and moving abyss of bodies, provoked by singular utterances and unspeakable pauses, a language which is the accomplice of silence, the guardian of secrets, a language which accompanies the most cherished rites of a group from birth to death, a language which—in immemorial myths as well as through the gesture of the writer—carries within in what Merleau-Ponty calls "the miracle of expressivity" (p. 91).

It is code-speak, a logic sequence or form made for and by machines and electronic assemblies. The networked evisceration of the sensuous materiality of place and position signal intensified degrees of alienation. It is one in which bodies and their voices are sucked up by surplus value in its incessant, global surging. The machine world and its code forms desymbolize the everyday. It is the command, the *cyber*-protocols that are charging human and animal movements. Codes without semantics, outside of temporality or location, a generic and would-be "transparency" of technization has participated in automating the rhythms of communal life and time. Thinking again along with Merleau-Ponty, in his continual mediations on language and expression, Janicaud declares, "thus the 'phantom of pure language' finds itself pushed to the extreme and realized: the myth of a communication without mystery, the 'algorithm' pursued by Western metaphysics" (p. 91).

The mutabilities of the subject and its cultural functions are always open to new folds, pliable within invented traditions and their purported futures. I identify the writing of these virtual futures as acts of *zoography*. I wish to avoid both extremes that delineate these futures: both a technophobic reflex on the one hand, and various emancipatory, and ultimately market-driven construals. These "zoographic" encounters require consideration of any particular organization of technical forces as transformative relations of power. It is what Paul Rabinow calls a "combat of forms," in the administration, maintenance, and control of populations, converging into historically specific forms of biopower.

In the context of his discussion of Rousseau's *Essays on Language*, Derrida employs the term *zoography* as it pertains to the myth of the origin of writing in the *Phaedrus*. He likens writing to a "painting of the living which stabilizes animality" (p. 292). A zoography is intended as a provisional marker for a way of thinking through the mimetic interplay of power circulating through specific new technologies that manage living and non-living systems alike. In Stiegler's reading of Leroi-Gourhan's sense of an originary technics, constitutive of human culture, "technological lineages are a relation of the human to matter analogous to zoological lineages" (p. 49). In this perspective, the technical is combinatory, like genetics or digital interfaces, a structuring simultaneously of matter and the "human" fabricator. *Homo faber* is continually, though not progressively constituted by the processes that configure social relations and subjectivity through technics.

As artificial memory systems themselves (like Plato's suspicion of writing) the technical also shapes the qualitative experience of temporality, its storage and computational networks extending cultural reserves. In the Paleolithic record, it is actualized as extensions of the anatomic and skeletal in the formation of flint chips. It is an exteriority that now unfolds the labyrinthine interiors of modernism's epistemic subject. Attending to Leroi-Gourhan's Paleolithic documentation of the relations between corticalization and technics, Stiegler regards the development of a distinctly human technics working through and expressive of a "mirror proto stage" in which "the differentiation of the cortex is determined by the tool just as much as that of the tool by the cortex: a mirror effect whereby one, looking at itself in the other is both deformed and formed in the process" (p. 158). Technoculture disquiets the partitions of liberal democratic states' grounding distinctions between the public and private, the discursive and the natural, interiority and exteriority and the corporate and corporeal. In Stiegler's reading "technics evolves more quickly than culture" (p. 15). As a force fostering "permanent innovation" technics regulates the languages of being and representation. The indistinction between life-forms and their management through the technics of biopower are "untimely" forays of a purportedly speculative theoretical curiosity. These artifices are, however, instrumental in their mastery over nature. Dressed in the rhetoric of progressive scientific development, technoculture, aggregates formerly disparate practices. I want to assert again that as practices of a quite distinctly "interested" theoretical inquiry they are catalytic generators for the control of cultural relations, institutions, knowledge production, and forms of subjectivity.

One of the key thematic threads of this project has been rethinking the nature of limits. It is a recurring theme, evident in its pertinences across sociohistorical structures, institutions, and forms of collective research. It recurs in multiple registers; as a question of the possible limits of knowledge attainable within the capacities of a material human finitude; as the continually contested border from which discovery, invention, and forays may be initiated, contained, or released from mythoreligious, theological, or sovereign forms of control; as the degrees to which science is itself limited by its peculiar rationality and methods.

The questions contemporary science asks and probes are as important as those it does and cannot, like the silences that make a melody take shape. The limits of scientific practice are limits to the quality of the realizations of its potentials. There are no predictable limits to what it is possible to theorize or to potentiate. The traffic between potentialization and making our theoretical curiosity appear are contingent upon economic, political, and ethical interests, as evident in controversies over NIH funding and the increased privatization of NASA projects. Rational choice debates appear to be sterilely devoid of care for embodied life, if not actually inclined to be mercenary. If techne produces artificial, if "useful" objects, what is its impact and relation to the making of theory as a material practice within the material world of our perception and relations? Academic science, at least in the United States, is no more autonomous than it ever pretended to be; it is dominated by techne, productive of utilitarian practices, its clinical and laboratory trials subject to predatory speculative interests in their potential for capitalization. Experimental scientific traditions, now outmoded but not epistemologically surpassed, are still contained by a logic of proof and verification, not a logic of discovery in service to the furtherance of a sustainable future.

Techne is the spinning disc, a mechanism of a persistence of the will to self-mastery. The terminal subject gives expression to what Nietzsche considered a "ruthless curiosity," an interlinking site for the new technologies of the self. Accumulating subject positions, the genomic and informatic subject remains in thrall to representation. Alterity and difference remain compelled by the logic of identity, the "proper" of non-contradiction. Fabricating the Same in globalizing digital or genetic informational flows the new paradigms and technics generate idealized phantasms in capitalism's own overproductive clonings and mutations. Generated in the increasing convergence of digital and genetic disciplinings of experience, and in compulsion to infinite accumulation, the split screen subject is as volatile as the digital market's own mutations. The genomic and informatic subject has traded its humanist shell, the individual commodity form of competitive individualism which it bids and sells in simulacral shares. For Nietzsche (1967), who wrote in the swelter of a Darwinian zeitgeist, a parallel question of value and relation prevailed in the manifestations of human mastery: "a new problem– today I should say that it was the problem of science itself, science considered for the first time as problematic, as questionable" (p. 18).

Zoographics

Zoographics is intended here as a portmanteau term for techniques, methods, and technical practices, such as transgenic agriculture, ex vitro reproductive technologies, DNA profiling, and somatic and cell line therapies. Through these varied and divergent practices of biopower daily life and identities are materially regulated, discursively maintained, and made normative in its patterned relations. The interiority of the acting human agent is displaced. The new genomic and informatic techne inflect distinctly discursive practices into everyday life. Their current ubiquity serves set standards, models, and the instrumental means variations on the humanist epistemic function of the subject. Genomic and informatic bodies are presented as plausible forms for a general cultural casting off of the individual as interiority. They are new twists in the skeins of modernity—the rewriting of the nature of embodiment that I am calling a zoography, or new loops in technics' tangled webs of the human and animal, nature and mind, matter and form. Genomic research, inscripted and encoded in informatics' signs, is one instance of zoography. It is a generalized economy of life processes and the differentiations of difference and becoming; genetic engineering is one actualization of a zoographic virtuality without origin or telos. Genomics and the informatic techne are one mode materializing the productive plenitude of technical tendencies; as a zoography, they are forms of writing and revision of genes in scripts for the management of a new "nature."

Reinscribing the domains by which a technological sovereignty is instantiated and resisted, zoographics addresses the biopolitical practices that discipline the continua of the animal, the inanimate, and the human. The forms of recognition for subjects, identity, and difference are being reconfigured. The representation of the human and its relations to the objects of its activity recode the normative, juridical supervision of life processes and experiences. In engineering untimely hybrids the nature and function of subjects are implicated. It may augur a hybrid naturalism sundering the legacy of Enlightenment reason and representation.

One unexpected site for a zoographic recoding has appeared in the work of "transgenic" artists. Alba, an albino rabbit, was born to glow phosphorescent green in the dark. With the assistance of geneticists, a fluorescent protein (EGP), a synthetic variant of a naturally existing gene expressed in the jellyfish Aquaria Victoria, was introduced through zygote microinjection. For Eduardo Kac, its designer, Alba exemplified a living instance of "transgenic art." For Kac the green fluorescent bunny (GFB) project was a "complex social event that starts with the creation of a chimerical animal that does not exist in nature." In his articulately philosophical review of the purpose and intent of his transgenic artwork projects, Kac listed several key objectives. Among them was an ongoing "contestation of the alleged supremacy of DNA in life creation in favor of a more complex understanding of the intertwined relationships between genetics, organism and environment," a challenge to notions of biological normalcy and viability, and a sharing and caring for "genetic material across traditional species barriers." Most important perhaps is his call for an "expansion of the present practical and conceptual boundaries of art making to incorporate life-invention" (2000, ekac.org/transgenic.index). Kac was at pains to clarify that his transgenic art is not a breeding project. He claimed that the innocuous genetic interventions he established in various projects allowed for interactive and dialogic relations to develop between species. He sponsored an aesthetic that would emphasize the "social rather than the formal aspects of life and biodiversity." Taking to task established notions of genetic purity, he encouraged projects that manifest and work within the "fluidity of the concept of species" (p. 8).[2]

While some of the artist-inventors involved in transgenic artworks, like Julian Laverdiere, give expression to a general cultural anxiety regarding the loss of the human and the potential for the unlimited production of monstrosities, Kac advocated the formation of transgenic social subjects. The domestication of creatures like Alba would also entail a concomitant domestication of humans. For Kac, "transgenic art is a mode of genetic inscription that is at once inside and outside the operational realm of molecular biology, negotiating the terrain between science and culture" (p. 8).

I wish to use Kac's singular exercises and play with genetic designs as an opportunity to briefly divert attention toward aesthetics. It is a point of departure for the kind of critical curiosity I have been advancing. The primacy of perception arises within the figures of the arts. In painting, color, line, depth, and form emerge within the movement of the artist's creative activity. A continuity and discontinuity in aesthetic practices can be found between projects like Kac's and those remarked upon by Merleau-Ponty (1964). As ways of "seeing," critical practices in the arts are saturated in the perceptual and sensible depths of theoretical practice. The classical divisions between the intelligible and the sensible, of abstract truth and phenomena, are broken down and reanimated. What I wish to point to here is whether or not an image can bring to appearance a presence, or the material imprint of an idea. For Slatman (2009) the ancient distinctions between pure essence, *eidos*, or idea, icon and images bear upon the means by which we have access to knowledge and participation in truths. Slatman works through some of Merleau-Ponty's insights into artistic and aesthetic practice, particularly with regard to the work of Cezanne, Matisse, Miro, and Paul Klee.

While the image has traditionally been relegated to the domain of the ephemeral and phenomenal, the icon transmits and relays simultaneously the presence and absence, and the visibility and invisibility of an idea, system of belief, or sustaining concept. They stand in contrast to icons which make visible what could not or was not permitted to be represented in sheer material forms. Contemporary visual and popular culture is dominated by idols. A "positive" ontology pervades social science and pedagogy. Attention and interest are contained within the voyeuristic gaze of the spectator. The persona of the *voyeur* "corresponds to the traditional idea of the spectator or the transcendental subject" (p. 201). She contrasts this unidirectional, controlling, and objectifying gaze with that of a Rimbaudian *voyant*, "someone who submits to the gaze of the other" in the reciprocal relations generated in the icon.

Transgenic art is only an instance of multiple sites of contemporary practice. It is a techne that plays and plics the tools—in their case, genetics, instruments, designs and protocols—that arrange everyday life and the means by which we increasingly live it. Transgenic art, like machine art or collages, work the concept of recombination. It reflects a cultural fascination with the increased convergence of engineering and biological models. They are hybrids reflecting the nature of what Rhineberger (2010) documents as the historical assemblages and "experimental systems" of molecular biology and its laboratory research. As aesthetic practices they are expressions of a distinct visual regime. Their materials and forms make statements that pertain to an era of representation that is made possible by the circulation and incorporation of their technical and epistemic objects. In the case of Kac, they are biological fabulations, recombinatory hybrids, and interspecies merging of the genetic resources.

Aesthetics was in no manner a derivative or reduced realm of critical interest for Merleau-Ponty (1997). For him any theoretical interest in painting was an attempt at metaphysics. In the making of art in Cezanne, Klee or in literature, particularly in Proust, he found a vital source for his attempt to develop a non-dualistic ontology that would place "expression" at the center of human practice. Expression is not subjective, as in Malraux. Languages are exemplary forms of expression; whether in speech, or in the "voices of silence" in the arts, it emerges within historical and intersubjective milieu. For Merleau-Ponty, Cartesian representation relinquishes all material resemblance between images and objects. The image, as ephemeral, curious phenomenon, is uprooted from its depths and horizons into the "natural light" of abstract mental images and representations. Invisibility, affect, and the virtual horizons of perception are exorcised. He writes that a Cartesian "does not see himself in the mirror; he sees a puppet, an 'outside.' His 'image' in the mirror is an effect of the mechanics of things" (1993, p. 131).

In his work, expression is a primordial and foundational surging through and between beings in the world. Transgenic art can be placed on a continuum of visual practices with the paintings of Cezanne that he found to signify a new depth and dimension within visibility. Cezanne's canvasses, in particular, made a rupture with the classical Euclidean and Renaissance practices of perspective. Linear perspective emerged from highly skilled techniques premised on a rationalist, calculative distancing of the artist and the objects of representation. It was itself a new regime of the visual in relation to classical and medieval models. The artist is absent from works that allow for a depth without a sense of moving, of lived bodies and objects in relation. Pertinent to some contemporary aesthetic practices, as in the work of Cy Twombly, Gerhardt Richter, and Francis Bacon, for instance, these are ways of seeing beyond the spectacle of the evident and immediately visible world of consumer culture. For Merleau-Ponty, Cezanne inaugurated painting as a form of expression that was carnal, embodied, and iconic. It is one in which the painter is as present as the painted. Quoting Paul Valery, the painter "takes his body with him" (p. 123). Slatman writes that in Merleau-Ponty's explorations into the perceptual world of art

> that it is precisely because painting arises within the reversibility of the flesh of the painter's body and the flesh of the visible world that there is an inner connection between painting and the world. The icon expresses this inner relation. The icon is the place where the looks are exchanged, the place of the chiasma of seeing and being seen and of the visible and the invisible (p. 200).

In his terms aesthetic practice is the pivotal chiasm that joins together the perceiver with fields of expressive perception. It is an aesthetic practice that permitted a multiplicity of perspectives of the lived body. It is a painting whose colors, textures, and lines of the moving hand are dimensions themselves; they open up a "second visibility" within material forms for the invisible to make its vivid appearances.

The increasingly slippery limits between the human and the animal, and life-forms from inanimate material substrates, raise the specter of the creation of our new monsters. Geneticist Francois Jacob (1973) described the emergence of cell theory in the context of early modernism's natural science, one in which the concern for mutations or "monsters" and their status has undergone change, becoming integrated into the practices of modern biological science. Jacob quotes Geoffroy Saint-Hillaire who said that "monstrousness is no longer a random disorder," (p. 124) an apt statement for the by-now well-established ability of removing and then replacing the nucleus of a cell with that of another species, as evidenced in the recent cow–human embryonic stem-cell disclosures. Craig Venter, never to be outdone, moved further by implanting an artificially created nucleus. In Jacob's internal history of biology "teratology, the study of monsters, was to provide biology with one of its main tools of analysis"(p. 124).

New genomic hybrids like Alba return like the repressed, unwanted golems and Frankensteins of the popular imagination. They are kin to the marvels and monstrosities of medieval bestiaries dazzlingly detailed with their dog-headed *cynocephali*, prodigious twins, or that of the *gryllus* with its neckless leonine head quizzically imposed on its feet. The new machinic beings making their untimely appearances are combinations of human and animal, of living forms with inorganic and technical prostheses. Nature and culture have always been hybrid and are now reflected in the complex of materials, organisms, and instruments of technosciences, particularly in molecular biology and genomics. Our new monsters are more likely to be installed as "up-grades" of the politics of visibility animating racism and class-marked differences. They are also likely to manifest in the first wave of somatic cell enhancements, appearing as gene altered realizations of perduring standards, attractive biomorphs, and postmodern renditions of the classic, athletic Nordic phenotypes. These were portrayed in one of the first films to worry these images. In *Gattaca* a future technological society is organized around genetically controlled classes. Gattaca's dystopia is a fully operative genomic culture. Class and race have effectively been transcoded into a genetic caste structure. The protagonist has managed to pass, enabling him full entitlement of the genetic brahmins who control all social functions and activities, including the vaunted space flight training academy he works in and where he aspires to be selected as a mission participant.

For Halberstam and Livingston (1985), the specter of the new hybrid monsters and goddesses is upon us, in making its appearances the

> rough beast that now slouches towards the next century is not monstrous simply by virtue of its status as a non-species: posthuman monstrosity and its bodily forms are recognizable because they occupy the overlap between the now and the then, the here and the always: the annunciation of posthumanity is always both premature and old news (p. 3).

In my zoographic conception (Zuss, 2000, 2003), symbiogenetic becomings constitute much of the evolutionary path of emergence and disappearance for molar organisms, including their "individuals," species and phyla. Let me return to the notion of the powers of curious thought as forces of attraction and repulsion. The "structural couplings" and folds of force within genomic and bioinformatic practices displace the intensive spaces, speeds, and partitions of the individual subject. It is a displacement and partitioning without model or origin. Technosciences and their objects of inquiry construct, shape, and alter one another. In Andrew Pickering's

(1995) terms, there is always some interplay or "dance" or "dialectic of resistance and accommodation" between human and "material agency" (p. 22). Machines, tools, and instruments have always been integral means for structuring human culture. The intentionality or desire of technoscience is always engaged in this "performative" dimension with and by the apparatus that instrumentalized its practice and the material world into which research is enmeshed. Despite Craig Venter's well-publicized interventions in implanting artificially produced cells with embryos, no existing ontogenetic or natural selective population diverts entirely from a phylogenetic or genotypic architecture. In the terminology of the practitioners of artificial life, it is a design with no original "self-description." This amounts to what might be considered in the philosophy of science as a "partial" realism.

The individual, conceptualized within a machinist or organicist notion of the body, stood aloof, in exterior relation to other bodies and the objects of its research. This objectifying, epistemological agent has regularly been challenged in modern thought. Merleau-Ponty's *chiasmus* is just one lucid expression of an image of relation that infuses interiority and exteriority within fields of perceptual sense and emergent knowledge. It eludes the trap of a dialectical opposition based on polarity or excluded middle. My sense of a critical phenomenological practice is one that accepts our already inseparable placement within fields of perception and historically accumulated patterns of thinking. To act and think curiously and imaginatively is nothing less than a fully embodied and interdisciplinary participation in the making and shaping of the depths and horizons, the outside of the inside, of any system of thought and experience. A critical phenomenological desire animates the question of embodiment and is carried in the collective circuits of attention in their iconic cloud chambers and scanning devices. Discerning the desires of theory begins to expel the scent of incense from the sanctuaries of subjectivity. It allows for thinking how the modern subject arrived fortified and propelled simultaneously by converging relations between political economies and technological instrumentation. It was born by distinct pedagogical practices fostering interiority.

Extracting its price, techne's genomic subjects are circuits of power and resistance. The genomic subject has formed another fold—its bioengineered bodies' networks of alternating currents of sexuality, crossing points for raced projects and their purported overcoming. Its future is commandeered by mostly privately financed, venture capital initiatives. Their sovereign rhetoric promises transformation, longevity, "sustainable agriculture" through transgenic crops and "immortal" cell lines. These disguise the continuity of humanist assumptions regarding identity and the management of difference. Its tropes cavort across a present charged with animals, electronic and mechanical prostheses, all extensions of humanism's outmoded embodiments. In one of the earlier and more nuanced renderings, by Halberstam and Livingston (1985), "the posthuman does not necessitate the obsolescence of the human; it does not represent an evolution or devolution of the human. Rather it participates in redistributions of difference and identity" (p. 10).

To think curiously is to participate in the co-emergence of assemblages, cooperating with forms of life and anorganic matter and energy that permit becomings, unpredictable arrangements and expressions of novelty to occur. The accent no longer falls on the first person, the syllable is de-stressed to include a symbiotic exchange that continues the process of what Foucault advocated in the creation a critical community, through a constant disengagement from the constituted modes of knowledge, category, representation, and identity. To think curiously is to partake of becoming, to open the future to think through us.

It is vital to consider the politics of genomic and informatic developments as they saturate new biosocial relations. Neither entirely novel nor unprecedented, they display a zoographic arrangement; they configure "us." It is a technical displacement of being. As an assemblage of institutions, researchers, and the biotechnical, it is deployed as a dominant mode of thinking and being. Theoretical curiosity is itself naturalized in the artificial reproductions of culture and its inhabitants. As Pearson (1997) has claimed

> the environment is an artificial world. There can be no return to a naive nature, and attempts to establish once and for all a natural order or balance on which to base an ethics or politics of technology is utterly foolish. There is only an excess of technics (p. 31).

Their untimely interventions propel and interrupt cultural expectations. Technics refigures the epistemic subject. It is the form "through which life lives itself," as Beardsworth (1988), Leroi-Gourhan (1993), and Steigler (1998) all propose. It is a perennial process of differentiation, adaptation, and reinvention of the material relations constitutive of cultural production. The acceleration of techniques for intervening into the tissue of daily life and the body, digital terminals and liminality of life processes, are productions of a deregulated, free-market industries. There is now an inextricable linkage between "natural" life, *zoe*, social cohesion, habitus, and the technosciences. This is particularly apparent in their influence on the medical and normative health models. These have become invested with the sovereign rights to regulate, define, and intervene in decisions over life and death. Informatic and genomic practices intervene, their discoveries germinating in a calculative curiosity founded in profit and property through the instrumentation of life.

In a similar vein, N. Katherine Hayles (1998) portrays the continuity of the human and the posthuman as a history that affirms a dematerialized cognition over embodiment. Hayles discerns the specificity of the "posthuman" in the development of technologies that prize "informational pattern over material instantiation." For Hayles

> the posthuman view configures human being so that it can be seamlessly articulated with intelligent machines. In the posthuman, there are no essential differences or absolute demarcations between bodily existence and computer simulation, cyber-netic mechanism and biological organism, robot teleology and human goals (p. 3).

This is a decidedly interested theoretical agenda. It is a rhetoric that portrays transethnic, transgender, transnational subjects as harbingers of a "posthuman" mode of existence. It would mark passing beyond the historical scarifications color and gender have imprinted on the surfaces of daily life. In this context, Stiegler's (1998) comment is apposite, that it is not useful to participate in "emptying the human of all specificity, but radically challenging the border between the animal and the human" (p. 136).

Genomic Sovereignty

The genomic is constructed of forms and values vitalized within contingent alliances, mergers, and interventions such as Venter's decoding machine in Celera's long, illusory race with Collins and Varmus at NIH for a complete mapping of the human genome. Embryonic and stem-cell development, in its very accessibility to researchers, remains in question. It is not unfair to claim that as the forms of life and life-forms are actualized in genomic, humanist, and reactionary values about the sanctity of life in general and the natural and the human come roaring to the surface. The instability of traditional humanist values are in part the effect already felt in the wake of genomic hybrids and reproductive choices and interventions. As increasingly disembodied forms are presented, nostalgic reclamations of the human person and its corporeality are intensifying. Biologic forms, often more virtual than actual, drive the exchange value of bodies, identities, and cultural representations.

Embryonic and stem-cell research, one of the most promising Orphic lines of theoretical practice in the biosciences, is confronting forces of reaction that would prefer the sacrifice of individual and collective health at the altar of their purportedly sacred pluripotent blastocysts. In pharmaceutical and germ-line interventions, there remains a crucial ambiguity for unanchored patterns of theoretical inquiry. Court decisions that may restrict NIH funding for laboratories with these cell lines show the confused encounter and public reception of certain paths of inquiry that are not and may never be clearly ameliorative. An unanchored and unaccountable theoretical practice faces the raging irrationality ablaze in fundamentalist systems of belief. Curiosity itself is aborted by evangelists.

Forms and values are uneasily negotiated in arrangements and compromises between ethical precepts and market strategies. Unstable, they pose problems regarding both the human and its engineered "posthuman" successor. Through the new technics new problematizations arise that ripple across the humanist plain. They pulse in the diastoles and systoles of a sovereign logocentric representation. Initiatives in research funding and institutional cooperation, as in the case of Caltech's ascendance in the early twentieth century or in Venter's Celera, genomic capital, adjusts the image and productivity of what had been the universal human subject in its forms of embodiment. Through hybridities of bacterial, insect and animal genes, organs and tissues, the heterogeneity of bodies is increasingly subject to disciplinary knowledges capable of regulating embodiment and dispersing former borders, generating singular expressions of difference.

A genomic sovereignty propels alliances and complicities of biopolitical practices—as a hegemonic, though disunified, competitive cluster of research agendas, funding arrangements, genomic and biotechnical investments incorporate the heterogeneous. Akin to the bacteria they frequently scrutinize, genomic capital entities constantly consume and reproduce each other. *Zoe*, the substratum of life processes, are discernable at the genetic level, bar-coded with market values in a representational, global economy as patentable forms, database, "HapMaps," and cell lines. Life-forms are sealed into archives established by the specific assemblages by which *bios* sets its meters and reads its vital signs. Genomic technics extend knowledges and powers over the body, releasing the energies of powers of bodies to act. Genomic research challenges different populations in particular ways. Its hybrid cell structures, constantly publicized, stir social anxiety in their challenge to what constitutes the properly and distinctly "natural" processes of conception, development, and expression.

The genomic paradigm participates in an ironic inversion of the grounding premises of the autonomous subject. As a subjectivity formed in an originary differential of power to forms of political and juridical sovereignty, the epistemic subject also existed apart from non-human "nature." Sovereignty, constituting the citizen-subjects of the emergent Republican states, also implicitly defined the extensions of human dominion over the natural world of animals and inorganic matter. Whether appearing in the rules of endogamy, immigration patterns, circumcision, or the availability of reproductive technologies, the contemporary exercise of sovereignty is increasingly a genetic and molecular biopolitics. This can be illustrated in the new recombinant somatic and stem-cell interventions that are being patented as inventions. Genomics participates in at least a partial sundering of the biological level of differences thought to comprise the distinctive species form of *Homo sapiens*. Its technics are now being given free rein in transgenic, xenoplant, and machinic interventions that alter humanism's balancing scale and upon which established biological and taxonomic distinctions are founded.[3]

The particular instantiation of a cluster of new relations, as in the cow–human embryonic cell, or in animal organ transplants, should not preclude consideration of the limits and connections established between technics, culture, and language in contemporary life. Despite the ascendant hyperbolic rhetoric emerging from "life science" multinationals like Monsanto, specific sites of intervention, interest, and limits to scientific competence are appearing; technoculture's "flyrooms," laboratories and arsenals realize historically particular expectations. They develop means for the control of a specific "biosocial" vision and can intervene radically into the shapes and performance of culture through its techniques for managing "our" possible futures.

Posthuman, hybrid and terminal subjects are inflected by digital and genomic grammars; they develop, erupting as forms of becoming from fashionings and fictionings of techne. Invented and maintained in specific, usually separate communities of practice, the posthuman is taking form out of the crafting of knowledges and power. These diverse and often competing communities of practice find themselves in thrall to finance capital. These forays—molecular, genetic, and informational—are diverted theoretical floodwaters. The new technics release an erotics of the singular, at once virtual, present, and vanishing; in calculable risks in service to "capital time," their currents shape, discard, and shift the substances of experience.

Technoculture can be considered itself as a new hybrid, productive of its own strange progeny. In making indeterminate the spaces and temporal relations of a posthuman modernism, it is a conjuncture of practices transgressing boundaries between the natural and the cultural, the constructed and the given, inherited world. Questioning the relations between the organic and inorganic matter stirs, while capitalizing from, a generalized cultural uncertainty about the nature of the human, the real and the eclipse of traditional truths. The problem or promise of unsettled boundaries and limits, discussed in the next chapter, is evident across techne's spectrum, both cellular and cyborgian. They do not obliterate modernism's imprint or the traces of its epistemological and axiological centers of practice. I propose that what needs to be examined now are the qualitative experiential features, aspects and limits of specific practices of, in a Deleuzian sense, becoming molecular and indeterminate, such as those starkly presented by germ-line therapies and cloning technologies. These genomic practices, in their alliance with informatics, deploy emergent forms of technicity that challenge and alter reproductive choices, the normativity of bodies, health, aging, and dying. I believe generative inquiries might most usefully examine the means, methods, and regulating alliances, theoretical and material, at work that comprise the rhetorics of the posthuman. As a contribution to a genealogy of technics, a zoography of the kind suggested here could appraise pragmatically how life is being learned and lived by other means.

Notes

1. Throughout this discussion I distinguish technics, as a generalized and differentiating component of material and expressive culture, from techne. Larry Hickman's *Dewey's Pragmatic Technology* (Indiana University Press, 1992) provides a useful definition of techne. In a Deweyan perspective, Hickman interprets the etymological origins of techne in Classical Greek thought to any productive skill or activity, differing as a human crafting of materials from activities generated by instinct or chance. Hickman writes "techne was thus used to designate a realm of activity that occupied a place between two extremes: the order of nature (or supernature) and the disorder of chance," (p. 17). He also shares Wolfgang Schadewald's description of techne as occupying an "intermediate place between mere experience or know-how, empeiria, and theoretical knowledge, episteme." See Mitcham, C. (1994). *Thinking through technology* (p. 116). Chicago, IL: University of Chicago Press.
2. Kac is joined by a growing number of artists contributing to critiques of genetic experimentation. In the "Paradise Lost and Found: Picturing the Genetic Revolution" exhibit held at Tang Teaching Museum in Sarasota, Springs, New York, artists like Bradley Rubenstein, Karl Mihail, Tran Kim-Trang, Eduardo Kac, Julian LaVerdiere, Heather Ackroyd and Dan Harvey continue an emergent aesthetic intervention into the technoscience of genetic engineering.
3. Kate Soper, arguing in the context of contemporary ecopolitics that it is critically "important that ecological argument avoids talking about the 'communality' of humans and animals in ways that conflate the biological and cultural and symbolic dimensions." In *What is nature?* (1995, Blackwell, p. 126).

Chapter 7
Minds, Limits, and Spaces

A contemporary expression of a will to unconstrained theoretical curiosity is evident across the spectrum of contemporary technoscientific research programs. It is what I identify as theoretical and material projects in their pursuit to overcome limits. The projects discussed in this chapter lay claim the right to an unimpeded disclosure of possibility through technical, genetic, and informatic research agendas. These projects authorize for themselves the challenge of the unbounded, of opening horizons and frontiers of research. Projects embarking on this "dare to know" trace the signature of modernity, as documented in earlier chapters, and find expression in the adventures of alchemists, physicists, and researchers, dating from Maupertuis to Gell-Mann and Venter. At the molecular scale of genes, the subatomic realm of quantum physics, or at astrophysical levels, this willingness to exceed and resist disciplinary and representational limits suffuses the hyperbolic, almost daily discoveries reported in the mainstream press. This is illustrated in the peculiar history of space flight and research, where a Promethean curiosity and interest is at stake. This was voiced by Daniel Goldin, the administrator of NASA, who stated boldly that "in many ways the campaign we are waging in Mars is similar to a military campaign." Goldin (2000), undeterred by the two repeated losses of Martian probes in the 1990s, grounded his campaigns in history. He advocated that in "heeding the lesson of George Danton in order to conquer, we need to dare, and always to dare" (p. 6).

As an intellectual adventure the trespass of codified knowledge and traditions can be an act of individual or institutional arrogance. A theoretical curiosity that claims impunity from any critical community grants its research immunity from ethical, public interests. Refusing public reflection and the philosophical questioning of it trajectories, it can force the management of objects of knowledge, whether natural and inanimate, material or conceptual, into instrumental coherence. Critiques of the abstracting processes of academic capital or knowledge work intend to play a subverting role in rechanneling the relations of knowledge and power performed through institutional support structures. Within a framework of informed public and secular interests, critical theoretical inquiry must be motivated to work as an ethical check to all final truth and knowledge claims.

A primary site for the rhetoric of surpassing limits has come from apostles of cyberculture and cognitive science. It represents the ascent of an unrestrained epistemic curiosity summoning images of a "posthuman" future. Richard Doyle

M. Zuss, *The Practice of Theoretical Curiosity*, Explorations of Educational Purpose 20, DOI 10.1007/978-94-007-2117-3_7, © Springer Science+Business Media B.V. 2012

traces Artificial Life's (AL) development from the Santa Fe Institute in the 1980s. Christopher Langton, one of its first exponents, stated that AL

> is the study of man-made systems that exhibit behaviors characteristic of natural living systems. It complements the traditional biological sciences concerned with the analysis of living organisms by attempting to synthesize life-like behaviors within computers and other artificial media (in Doyle, 1998, p. 306).

As a strongly funded research program dedicated to fabricating lifelike behaviors, AL's ethos manifests a renewed vitalism in attempting to delineate the origins and essences of life processes and intelligence. Researchers like Langton, Moravec, Steels, and Minsky mark a departure from Norbert Wiener's first-generation cybernetic culture. At its inception, researchers were committed to the central dogma of the "code-script," "program," or informational processing nature of machine and organism. Langston's generation seeks signs of "second order emergence," the capacity of complex systems to alter, modify, and develop their own programs. N. Katherine Hayles (1998) distinguishes two generations of research programs and practices by differences:

> first-order emergence denotes any properties that are generated by interactions between components, that is, properties that emerge as a result of those interactions, in contrast to properties inherent in the components themselves. Among all such emergent properties, second-order emergence grants special privilege to those that bestow additional functionality on the system (p. 243).

AL researchers lay claim to non-carbon-based manifestations of life-forms. In the biosciences these are now defined as self-organizing and replicating systems with the capacity to evolve. In a review of AL research, Claus Emmeche (1991) stated that "complex-system research is at risk of being driven into an advanced form of essentialist thinking with its continual assertions that life is a collective property in complex self-organized systems that can emerge from many media" (p. 165). Doyle, Emmeche, and Hayles hear an echo of Stahl's nineteenth-century vitalism at work in the cutting-edge simulations and models of AL laboratories. The attempt to synthesize, simulate, and digitally replicate the core features that make life possible are evident in AL's statements of purpose. For Doyle, "A life seeks to derive the formal nature of the living system, life's algorithm, by abstracting it from its material, carbon-based prison" (p. 316). Whereas nineteenth-century vitalists sought to isolate and distill an imaginary vital force, the reduced essential component of a system, AL researchers have attempted to synthesize and build on the structural elements of life-like processes, in which it could become possible, as Hayles claims, to "procreate by emergence" (p. 234). Doyle quotes Langton, who, like Hans Moravec, would have us "download" consciousness onto a disk. Doyle believes that "life, as a physical process, could 'haunt' other physical material. AL can contribute to theoretical biology by locating life-as-we-know-it within the larger picture of life-as-it-could-be." Doyle recognizes this "rhetoric of computational vitality" as more than hyperbolic self-promotion and speculation. In his analysis, "it is a dream grounded in the history of automata and life, a dream based on the scientific desire to 'know what life is' " (p. 309).

Among the myriad projects in progress since the 1990s have been attempts at the "genetic programming" of human embodiment. Software has been designed by several laboratories that imitate the complex processes within cell development. With a specific objective of finding the actual gene interactions at work in normal and pathological conditions, projects like Stanford's "E-cell," or the "cybernetic cell" programs produced functional models of actual cell activity. The Entelos project was created as a working model of a fat-cell to mimic how diabetic and obese patients would respond to diet and drug therapies and of much interest to pharmaceutical giants Eli Lilly, Bristol-Myers, Squibb, and Johnson & Johnson. Researchers asserted the importance of biological "robustness" at the center of their investigations. Referring to the ability of biological systems to adapt to changes in temperature, the supply of food, the intrusion of toxins and chemicals, the notion of robustness presumes that all physiological systems, in general, and cells in particular, must have "backup systems" that allow them to adjust to internal and external forms of interference. It presented a new method for observing adaptive mechanisms. B. Pallson, who headed the genetic circuits group at U.C. San Diego claimed that cybernetic cell research indicated that "we're witnessing a grand-scale Kuhnian revolution in biology" (in Gibbs, 2001, p. 53). The "E-cell" model itself built from software that mimicked genes was a the basis for the research team of the Laboratory for Bioinformatics at Keio University in Fujisawa, Japan, where they were intent upon the extremely reductive process of disclosing the "minimal number of genes needed to create a self-sufficient organism and then synthesize it"(Gibbs, p. 55).

Testing the limits of theoretical curiosity bioinformatics also invents new mixtures of the living and the artificial. Research has been directed toward the incorporation of alternate pairs of amino acids for incorporation into the proteins comprising the DNA molecule. Utilizing a variety of amino acids in combination with all living systems' structure of four bases, researchers like Eric Kool at Stanford could make the claim, in 2001, that in "five to ten years, we'll have an alien replicating system." This research required the bonding into pairs of the anomalous amino acid, either using methods of generalized substitution or the placing of new amino acids at specific locations. These accelerated initiatives were sponsored by the Genomics Institute of the Novartis Research Foundation, California Institute of Technology, and Stanford University. These projects have been avidly interested in the viability of organisms structured with qualitatively different genetic codes. "Scientists say creatures with a truly different genetic code would essentially be alien life forms," Philip Hilts (2001) reported. Suggesting connections to issues arising from astrobiology discussed in this chapter, Hilts found that "one of the aims of the research is to see what kinds of life may be possible outside earth" (p. 2).

Some of the most vigorous defense and critique of a purported "posthumanity" are present in recent feminist criticism. For Donna Haraway and Sherry Turkle (1995) the hybrid interfaces of emergent cyberspace identities permit unprecedented relational matrices to take shape. In digitally liminal interactions, gender is performed. Gender identity is radically contextual, comprised by both fusions of virtual and "real" or material social bodies. The "intimate machine" of Turkle's "culture of simulation" purposely blurs all boundaries of gendered embodiment. Within the synchronized social relations of knowledge and the exchange of information social networked, hypertexted sexual personae enjoy a freedom of expression unavailable in RL. For Turkle, a feminist hybrid subject propels women within a realm of positive freedom, in relations of power—the new circuits of desire of digital transfer. The cyborg woman subject compels agency in the enabling conditions provided by a technology of the terminal subject. She informs new relational potentials and possibilities released from modernism's scripts of bounded gender relations and representations.

Lee Quinby (1997) ably critiques the cyborgian model idealized by Haraway and Turkle. She questions the dematerialization and abstraction at work in their fabrication. Claiming that all cyborgs are not born the same, Quinby worries over Haraway's "flattening out of subjectivity into the cyborg and her full-scale slide from metaphor to ontology," one in which we are made to expect that "all cyborgs were the same and all hypertext formations cast a single shadow" (p. 229). She accepts the possibility of body-machine identities as presenting a problematization of the oppressive and instrumental interests of technoculture for feminism. Quinby, however, does not accept that cybertechnologies will vitiate embodiment. The terminal subject of virtual reality and cyborg identities expresses a foundational hostility to gendered bodies:

> from the Book of Revelation to the Heaven's Gate website, denial of embodiment has been a heterosexist obsession that defines itself oppositionally to women's bodily excess and lesbian and gay sexuality (p. 236).

The distinctions between material and immaterial embodiment, living and non-living, and human and non-human systems exponentially expand the reach of second-order researcher beyond their first-order AL predecessors. The parallels constituted earlier in molecular biology, genetics, and information, between material substrates and energy, programs and codes take on a dramatic qualitative shift. To Doyle it is "evaporation of the difference between living and non-living systems. Once housed in an invisible unity, then a 'secret,' life now finds itself without an address." He pays tribute to Ed Regis, who unequivocally stated that "we would like to build models that are so lifelike that they would cease to be models of life and become examples of life themselves" (pp. 317–318). This kind of hyperbole has become common coin among the advocates of a "transhuman" future. It has formed the core curricula for Singularity University which offers a syllabus that includes seminars on the "end of aging" and achieving physical immortality. Ray Kurzweil, an engineer and leader of the school, expects an unfettered acceleration of technologies in everyday life and relations. In Philippe Riviere's (2010) review, Kurzweil's vision is of humanity soon coming to a "point ('the Singularity') where technology will saturate the universe with an intelligence independent of its biological origins" (p. 14).

Among the problems raised by current complex-systems thinking in biology and genomics research practices alike, are questions of whether any privilege remains intact for organic material and their forms and relations over inorganic matter. The inviolability of forms of life interventions conjoins these traditionally separated domains of knowledge, identity, and practice. Abiding concern for "partial," direct and indirect realism in philosophy of science requires reconsideration of the interplay and human and "material agency."[1] This includes consideration of forms of life and life-forms in terms of "collective properties," whether as self-organizing, complex, or autopoietic systems. The suggestions of autopoietic theory stress organization, closed structures, and relative autonomy. For Niklas Luhman (1995) and Cary Wolfe (1998), for instance, they are delineated by a precise "cut" or fold that situates and enacts the development of a self-reproducing system. They tend to emphasize homeostasis, repetition, order.[2] Here, contra Maturana and Varela (1980), I wish to stress differentiation and the becoming—molecular, life-systems as enclosures and multiple expressions and instantiations of intensities. They are configurations of a virtual organization never fully actualized. Among the many questions for this technoscientific pedagogy in general and to the field of genomics in particular, are the following: does this enterprise expedite the self-organizing quality of emergent systems? does it serve as a catalyst for an inherent "second-order" generativity in natural and artificial systems alike? In the framework of this inquiry, genomics is a techne of dispersive or rhizomatic complexity. As I discussed it in the larger context of technics in Chapter 6, it is one in which the modernist human subject becomes a mobile, indeterminate site. Embodiment and the processes of individuation long pursued in the natural sciences become fragmented participants within moving ensembles, assemblages or clusters of life-forms, forms of life, and self-organizing matter.

I wish to elaborate on the distinctive features of a genomic-based technics. As one, still dominant expression of a perennial virtuality, genomic and bioinformatic practices imply significant reformulations of modernist epistemological values. Their implications concern extensions as well as departures, alliances and tensions between rival research programs within the very competitive heterogeneity of technoculture. As discussed earlier, they pose crucial questions regarding scientific practice and philosophical tenets of embodiment, interiority, and materiality. Four general distinctions marked by a genomic or "zooetechnic" episteme constitutes its challenges to some of the basic and traditional criteria for knowledge as well as the directions for theoretical curiosity to take, as transmitted in cultural practices and science education. First is its challenge regarding the pliability and limits of the relations of form and matter. This issue makes for a genetically inscribed reappearance of classical hylomorphist questions and expressions of vitalism. Among these are the relative degree of independence and causality generated by form, pattern, and structure. As efficient causes, genomic practices, simulations, and models are autopoietic, self-generating, and replicating forms of information.[3] This marks its second distinctive feature. Genomic interventions parallel, but are irreducible to the productive acceleration of capital flows. This opens an opportunity to consider the degree of abstraction and autonomy of "information" over energy and material embodiment desired. At what point and how are decisions made in response to the potential mutability of genes, cells, tissues, organs, reproductive processes, prostheses, and "life-forms"?[4] Codes and capital function as general equivalents. One illustration is emerging in the field of embryogenesis. Melinda Cooper (2008) has compellingly documented methods adapted from speculative topology, particularly in the tissue engineering of embryos, which could permit an endless "surplus" of life-forms.

For molecular biologists and embryologists a reductive level of analysis explains the function of genes and their relations in cellular growth and development. Researchers have often sponsored a public discourse that attempts to explain cellular and organismic expressions of behavior, disease, and health at the unit level of genes. At the same time, genomics, under the spell of computational and complexity models, attempts a false holism or synthetic explanation of living processes. This is readily heard in the hyperbolic clichés of master codes and secrets to living processes that welcome back a variant of vitalism humming in Celera's genome decoder machines. As an effect of these three distinctive features, the fourth issue remains the problematic nature of embodiment, health, and reproduction. At the present, the possibilities and potentials for intervention—whether at the somatic, germ line, tissue level, as developing in research (and irrational reaction) on embryonic stem cells, as well as the telomerase edges of DNA that might extend longevity—remain within the control of a technocultural elite. Private enterprise maintains almost sovereign control over the intellectual property, patents, and the development of forms of embodiment and corporeality. Among these are prospectively valuable genes, DNA replication processes like PCR, protein unfolding imaging, and germ-line therapies.

The intentional generation of hybrid life-forms are expressions of a free market will to theoretical curiosity. As evolutionary responses to environmental, ecological milieu, life-forms in a "natural" state are always temporally embodied relations and structures, continually subject to new perturbations. Self-reproducing, complex, and adaptive, biological embodiment requires continuity and coherence within divergent and changing environmental influences. While discontinuities do importantly occur in evolutionary adaptation, they function within the material structures and organic processes of species, not always subject to instrumental and hypothetical complexity models.

Figuring Futures

Revisiting the question of the principles of generation of natural, living things, the place of the biosciences today can be brought into critical focus. In Aristotle's *De Anima* nature, as *physis*, is grounded by an inherent principle, an a priori that makes manifest all its actual life-forms, including the variety and changes in their growth, maturation, and decline. For Aristotle this a priori, is the generative potential principle of motion. The power to move of its accord is the sign of a natural life. This movement or *kinesis*, along with its contrary, rest, is the motivating force of the psyche in its bonding with the material matter it animates. At a metaphysical level, one that is not separate from its material actualizations, the psyche itself is immanent to and participates in the ontological substance of being and becoming.

While it is not within the compass of my competence to bring into consideration of the questions of being and substance, questions of first principles like these have marked a continuum of theoretical curiosity from Thales to Heidegger. They are questions securely within the realm of a purely speculative theoretical and intellectual curiosity. In the following I wish to contain the range of this discussion to the production of practices in the technosciences and its pedagogies. Ontological concerns should, however, remain in reserve as the often-unstated grounds not only for any implicit claims to transhistorical truth or reason, but also as the unquestioned, idealized motivators legitimizing trajectories of social research. The implications of human agents intervening within its own nature and the natural world are core issues for any reflection on the directions of the apparently "insatiable curiosity" of contemporary technoscience.

The issue of techne and technics remains central to this reflection. Among the many critical questions that might be raised are the following: How is techne, as the ability of institutional agents to design and plan specific theoretical and material objects, set into motion? What is the dynamic that generates artificial beings, whether through genomics, neuroscience, or artificial life? If there is a generative principle to life-forms, as the activating kinetic motivator of their existence, what is the principle and nature of things created through technics? And do we belong any longer (if ever there was a pristine state prior to techne), in the unfolding of a limitless physis?

We are engaged in a matrix of natural and artificial lives. Hybrid mutations and simulations rush across laboratory screens. They present a fundamental mediation of the experience of embodied existence. How are we to figure a phenomenal, experiential sense of place for these complex emergent relations and for inhabitable futures? The nature of the "natural" has always been a slippery notion, evading closure as human work, tools, and techniques have broken the very ground of the natural world to eke out its nurture.

From the Aristotelian tradition, potentiality is paramount here. It is given the status of a founding or underlying ground, the *arche*, for the actualized movements of all natural life. Instead of the elements or mixings of the primary elements and colors that the Presocratics debated, potentiality represents the generative principle of nature itself. If we are to accept this principle of formation, the following questions might be asked: how is technoscience joined to nature? Are nature and techne allies or opponents in the theoretical projects of contemporary science? Techne signifies the ability to make things appear that do not in themselves have the principle of motion. The manufacture of things that do not exist in nature brings into question the limits of the potentiality at work in the practice of theoretical lines of inquiry that themselves are not often guided by any first principles. Techne concerns itself with the practical crafting, design, and planning of ideas, images, and concepts in order to produce things and beings. The limits of theoretical curiosity, in its institutional forms, particularly in the thicket of start-up, venture capital enterprises, return in the production of techniques into and between bodies, as well as the surplus of novel forms of life that they may make appear.

In what came to be called Aristotle's "natural books," particularly in the *De Anima*, inquiry into the specific question regarding the developmental process and nature of life emerged. Among the many, perhaps unresolvable, speculative problems probed is the question of the relation between the soul or psyche and the body. He proposed a twofold concatenation of the dynamics of the *pysche,* or soul, as source and motivating force, at once in relation to and inseparable from corporeal existence. While this relation certainly formed the basis for theological, hermetic, and protoscientific speculations, it retains value for this project in raising the question of the enigma of what life is and how and what it is becoming. For Aristotle, while life, made manifest through and by the psyche, is an a priori principle of all nature, it is itself without essence, hollow and abstract. Eugene Thacker (2009) comments that for Aristotle, in the *De Anima*, the principle of Life, or that which conditions the living, is absolutely non-existent. Life and the living occupy entirely distinct epistemological places. Thacker notes that Aristotle states clearly that there is "no such thing as Life in itself—only Life as it is manifested and actualized. Life is an emptiness that accounts for the fullness of the living" (p. 36).

Life is a disembodied principle. In our contemporary deployments of the technosciences, especially in the potential influence and proliferation of life sciences, a questioning of the nature of life is a critical concern. A biotechnics now intervenes across former boundaries between the animate and inanimate. Its "experimental systems," as Rhineberger (2010) documents, since the 1960s have included an imaging armamentarium through electron microscopy, chromatography, X-ray structure analysis, radioactive, liquid scintillation counting, and electrophoresis. These are instances of the means for a zoographic inscription instances of "life" through experimentation in "laboratory life." The theoretical forays of Aristotle are germane to rising to the challenge of how, whether and in what situations, we want and can act upon a life regarded as calculable. Is life subject to the Promethean attitude, as a managed, technical object open to design? Nothing less than embodied life, in its multiplicity, is now in question. The place of the human is suspended and indeterminate; in phenomenological terms, the very experience of living is itself a theme to be "bracketed," open for critical reflection.

In the classical context discussed in the first chapter, pedagogies of identity and posthumanity present a host of potential dynamic relations that are coming to bear upon the quality of the everyday. For Heidegger (1977), contemporary sciences operate under the spell of a technological dispersion, the effect of a boundless ontological Being radically diminished into "produced" beings and entities. Among the questions that are raised for this project, premised on constant reconsideration of the experiential quality and sense of theoretical practice are: What shapes are we making to the experience of life? Will the new interdisciplinary sciences of life open horizons of potentiality for sustainable and ecologically adaptable forms of embodiment? Are we learning to live amid renewed relations that will foster what Merleau-Ponty called "wild being" and becoming? It is a future that might not remain familiarly human, that indeed may be now in the process be claimed to be "posthuman." An open future without humanity at the center may pivot on the dangerous promise of an otherness captured by a technics of life. The twenty-first century, as it confronts or denies imminent ecological crises, also must come to terms with its technical prowess. The technosciences in their theoretical practice and potential have amassed the global means of production of the virtual, a surplus of generativity.

Theoretical curiosity is never a closed system. It is rhizomatic branching in its open-endedness and lack of finality. Fields as divergent as neurophilosophy, complex systems biology, or quantum physics are assemblages, composed of multiple, intersecting technical-cultural systems, apparatuses, instruments, and organisms (including the researchers) and ways of interacting with the phenomenal world. Theoretical curiosity participates in these symbiotic systems, its vitality composed of intensities, differentials of difference as the only ground. To say that curiosity is assembled, is to state its flux and transversal across contingent historical linkages and situated networks that form and dissolve into other temporal and spatial unities. It is also to place the vitality of curiosity as a crossing point between human and non-human agents, to interpose the animate and the inanimate, to recognize the monstrous in the familiar. In this vertigo, both epistemic and ontological, the making of the "we" of nations and communities may be effaced. In the language of technics at least since the advent of the machinic model of Descartes and la Mettrie, the accent here is on a dramatic shedding of anthropomorphism. The face of the human, opposed to the animal and "natural," has carried through a whole epistemic reign in the natural sciences. It is an eclipse of traditional humanism, in its representational and non-emergent, causal models, axioms and algorithms. The technics at hand call for a posthumanity of emergent patterns and flows.

A figure like Maupertuis, portrayed in Chapter 2, is apt to considering a will to power radiating from and around theoretical desire. As a relentless pursuer of new frontiers of knowledge, this eighteenth-century natural scientist prefigured developments now at the cusp of genetic technologies (discussed in Chapter 6). His pursuits, crossing physical, moral, and epistemological barriers, also resonate within contemporary debates regarding the limits of knowledge with regard to the technosciences. Hans Blumenberg (1985) comments that for Maupertuis the work of curiosity manifests a "constitutional condition of a being that is no longer able to see its original connections, that collides with its de facto locatedness" (p. 415). It is an epistemic and ethical dislocation. His trespasses act as a signature gesture of epistemic violence, an ungrounding not only of the objects of research but of its subject as well. Maupertuis's will to system, his flagrant disregard for moral and mortal finitude in attempting to build his system of "genetic totality," echoes throughout modernism's rarefied epistemological obsessions with the bases and validation of knowledge and meaning, as evident in the dominant Anglo-American philosophy of language and mind. Like other natural scientists beginning with Francis Bacon, Maupertuis dispatches interpretive emissaries into an imagined universal past as well as into the progressive unfolding of an absolute future.

He is a figure for the abstract mathematical labor, as observed by Sohn-Rethel (1977) and Husserl (1970), among others, one who has authorized his projects in complete detachment from embodied and relational identity. It is a "constitutional condition" no longer overtly dependent and connected to any living subjects and their lifeworlds. Parallel with systems of monetary exchange, it is severed, while capitalizing from practices of human activity and inquiry. An autonomous mode of inquiry, its epistemic and often quite physical violence is imposed on its objects and subjects. Intellectual and basic research labors of this kind, as illustrated in the massive Hadron collider inside an excavated Swiss mountain, are increasingly endemic to an instrumental ethos in the natural sciences. They often function within internal formal protocols, algorithms and market regulated "just-in-time" efficiency criteria. Its self-enclosed, often proprietary, models and systems pursue quantifiable validity, predictive and deductive logics, acting without apparent relation or response to other spheres of material and cultural reproduction.

Figures since Maupertuis, who could include contemporaries like Oppenheimer, Gell-Mann, and Feynmann in physics, or Venter, Moravec, and Doyle in genomics and bioinformatics, participate in an economy of inquiry shaped by and for the material objects, tools, and instruments by which they derive their models and representations. This dislocation of spatial and temporal orders need not be regarded as trespassing across the integrity of traditional practices or methods. System builders of modernity, such as not only Bacon, Descartes, Leibniz, Kant, and Hegel, but also Linnaeus, Ramus, and Buffon, in the natural sciences, for instance, purify the exceptional and the divergent for their unifying projects. An instance of these distinctively synoptic practices was illustrated in one methodical consideration of the anomalous and monstrous during the emergence of Renaissance natural science. The *Lincei* Academy of Naturalists, which included friends of Galileo, convened in April 1625 to dissect and analyze a two-headed calf. They acted in the interest (and in the presence of the Pope Urban VII) in the interest of disclosing the principles of organization and "common character" of the species as a whole. This case study of an anomalous mutation was the occasion for deriving general structural principles while marking its divergence from the normal reproduction. An older reading of the "signatures" of nature, in what Ginzburg (1983, p. 99) calls a "conjunctural" paradigm, and in what later C. S. Peirce defined as an "abductive" logic, are replaced. The particular and specific differences were put under censure and catalogued. Difference was effaced within an overarching explanatory causal model. It illustrates a historical turn and crossing point in method, model, and epistemic pursuits. The interest in patterns, signs, and differences were becoming replaced by the taxonomies and schema of emerging paradigms that generalized and "levelled" or rounded off from the particular, the contingent and anomalous phenomena.

Functioning within relatively closed economies of thought, current pursuits of unifying "theories of everything" like Kurzweil's "Singularity" (with its university) correspond with the abstracting systems by which surplus value and wealth have always been extracted from actual physical and intellectual productive labor. System builders depart from the work of any collective inquiry in the teleological, deductive models they construe. Projects like Gell-Mann's pioneering search for an explanatory basis for phenomena and structure in subatomic, nuclear particles, for instance, continues, despite its novel findings and instrumental means, a tradition committed to causal mechanisms sufficient in themselves to define and represent aspects of the natural world. Gell-Mann's Prometheanism is expressive of the continuing social valorization of an unlimited epistemic curiosity. Gell-Mann was a pivotal leader in the search for the basic building blocks and forces of nuclear physics. A colleague of such pioneers in subatomic experiments as Fermi and Oppenheimer, Gell-Mann is credited with "discovery" of the quark and the principles underlying what would develop as a unified field theory. Chet Raymo, the reviewer of George Johnson's *Strange Beauty*: *Murray Gell-Mann and the Revolution in Twentieth Century Physics,* a physicist himself, records what Gell-Mann, in his Nobel address asserted: "we are driven by the insatiable curiosity of the scientist, and our work is a delightful game. I am frequently astonished that it so often results in correct predictions of experimental results." Raymo (2000) concludes that "Johnson's strangely beautiful book captures the curiosity, the game and the astonishment" (p. 101).

Avatars, like Maupertuis in modern science, prefigure Watson and Crick, Craig Venter, Gell-Mann, and Hans Moravec in the twentieth century. They are agents of a purposeful, sometimes ruthless theoretical curiosity. Their empires of knowledge have exerted the power to affect the growth and quality of understanding across disciplines. They also act, ironically, to bring into relief certain unexpected configurations. Adventurers in theory seize historical moments that offer up apertures for transgressions which in our day are frequently sponsored and accredited by aggressively market-driven interest in developing interdisciplinarity. Academic capital is appropriated, patenting what appears to be singular. Akin to the figure of Hume's hunter or gamer in pursuit of knowledge, institutionalized inquiries are goaded, directed toward some certain utility. Entrepreneurs of the mind, speculative theoretical work is contracted and outsourced to laboratories in profitable "start-up" experimental research and development companies in genomics, AL, and other new mergers of technoscientific expertise. Their hunting and gaming take place in volatile competitive global rivalries and market uncertainties. Kaushik Sunder Rajan (2006) provides a rich ethnographic study of this hunting in drug development and genomic research in Indian and US venture capital companies. The much publicized history of the Genome Project illustrated the competition set into motion between Francis Collins and the National Institutes of Health with Craig Venter's Celera decoding instruments. Once again, Maupertuis, more than a straw man for the would-be "hero" of curiosity, represents the continuing danger present in advocates of an unbridled and privatized "pure" speculative curiosity. Its "blue-sky" veneer dissimulates an appropriating epistemological will to power. It is a contest over forms and figures and systems of life, including their animal, sometimes human, bodies and genetic structures. The privatization of life-forms, including the ability to patent as "inventions" organs like the spleen, the umbilical cord and retinal tissue, for instance, places modern theoretical pirates beyond and outside the pale of intellectual desire. Maupertuis's limit case exemplifies the uncertain territories arrived at when primary ontological, ethical, and pedagogical grounds remain deeply unsettled.

In the genomic and artificial life spheres of research, to think curiously is to participate in the co-emergence of assemblages. Operating or cooperating with forms of life, an organic matter and energy permits unpredictable arrangements and expressions of novelty to occur. Their "experimental systems" are hybrids themselves. As Tim Lenoir writes

> the experimental systems that molecular biologists design are "future generating machines," configurations of experimental apparatus, techniques, layers of tacit knowledge, and inscription devices for creating sem-stable environments—little pockets of controlled chaos—just sufficient to engender unprecedented, surprising events (in Rhineberger, 2010, xiv).

The accent no longer falls of the first person. The reflexive human "I" syllable is de-stressed to augment symbiotic exchanges. Theoretical curiosity remains a potentiality. Its valences are multiple, "emergent," and must remain open to contest in public, democratic, and popular engagements. Heterogeneous and convergent technoscientific cultures pose challenges to the nature of collective, ecological experience. They may come to reinvigorate critical communities of practice, as in the "practices of self" Foucault advocated in a disciplined and constant pedagogical disengagement from the constituted modes of knowledge, category, representation, and identity. To work, act, and imagine curiously then begins to decide upon certain collective horizons of becoming and the possible, to open the future to think through us.

Martian Interlude

The rhetoric of surpassing limits prevails in biological and physical knowledge of processes subatomic, both genetic and galactic. Astrophysical knowledge secured to an unprecedented degree, with projects like the rejuvenated Hubble telescope or the space stations and highly sophisticated observatories like Keck, allows for detection of events and formations at distances and temporal scales approximating the limit of the materially detectable universe. Instruments and interventions are being pursued that operate in previously unimaginably small and large scales of time and space. Nanotechnologies are emerging that permit stopping the motion of electrons. The "*attoseond*" field, for instance, starting out as a purely speculative basic research theoretical pursuit like the Hadron Collider, works within the time scale of a billionth of a billionth of a second.[5] The technics at work in the decoding and sequencing of the human genome were part of the general augmentation of instruments, whether informatic, genomic, or astrophysical, which are directed toward disclosing what are claimed to be founding processes, laws, codes, and dynamics.

A recurring theme in the investigations of the forms and processes at work, whether subatomic or cosmological, is the question of potentiality. It informs interdisciplinary research tendencies, shared by the otherwise-heterogeneous subcultures of the technosciences I am reviewing here. It is largely research concerned with the potentiality of systems, whether molecular, cellular, genomic, or in quanta of particle, atomic, and even neurobiological levels of analysis. Common to these diverse researches and their imaging technologies is an abiding interest in the relation, transfer, and "emergent" reciprocities between magnitudes of energy and their instantiations in mutable forms of matter.

This theoretical interest motivates the dramatic pursuit of origins of life, both terrestrial and extraterrestrial. This interest is manifest in sustained exploration projects, as in Search for Extraterrestial Intelligence (SETI) and organizations such as the Planetary Society. The persisting notion that the potential physical, chemical, and environmental factors necessary for life must be, at least occasionally, realized and coalesced, drives some of the specific projects of NASA and the European Space Agency. Shapiro and Greenberg (1998) satirize the debate over the forms of life debate by contrasting what they identify as advocates of a "*predestinist*" outlook to a "*carboquist*" orientation. The predestinists argue that with the abundance of light elements, including amino acids, purines, pyrimidines, and sugars, constitutive of nucleic acids as well as proteins, "life everywhere will be based not only on carbon chemistry, but on carbon chemistry similar to (although not identical with) our own" (p. 254). The carboquist camp, on the other hand, arguing for a variant of what Stephen Jay Gould identified as the "anthropic" principle, claim that life elsewhere in the universe must resemble that present on Earth, because no other basis for life can exist. Our own chemical system, particularly in its use of water, as a solvent, and carbon, as the key building block of large molecules, is uniquely fit for the purpose of sustaining life (p. 188).

The question of an a priori natural "fitness" is congenial to those who adopt variants of the anthropic principle or what Freeman Dyson calls "animism." Gould regards it as "the idea that intelligent life lies foreshadowed in the laws of nature and the structure of the universe" (in Leslie, p. 188). It is an idea that posits a prefiguring of life-forms, and its rudiments, for both simple and complex, intelligent manifestations. It has found advocates from early twentieth-century proponents from Lawrence Henderson to Dyson, a popular contemporary physicist and critic.

Thinking the limits of a creative force and potentiality has been a perennial concern. Crossing historical and epistemological borders, it has fortified doctrine, practice, and ritual in all the world's varied divinatory, suprahuman, or immanent manifestations and avatars. One well-traveled way of questioning the nature of potential and the possible has been variously articulated as a questioning of plenitude. It is a controversy, debate or corpus of beliefs that ponder the productivity of the universe, both material and immaterial.

As an intellectual adventure, the trespass of codified knowledge in artificial life, astrobiology and space explorations can be dressed up as theoretical speculation on first principles of nature and foundational forces at work in physical and biological processes. In his study of the scientific bases for forms of life and a "plurality of worlds," Steven Dick (1998) documents a long history reliant on its essentially theological premised plenitude. Dating variations of this notion to Epicurus and Lucretius, the principle of plenitude holds that "when abundant matter is ready, when space is on hand, and no thing and no cause hinders, things must assuredly be done and completed" (p. 12). It is a claim for a foundational procreative abundance of nature in all its possible manifestations.

In its variant practices and research programs over the centuries and into our time it is the search for the pedagogic and unifying generative principle of an originary agent, whether god-like, causal laws, a "ghost in the machine," or in "emergence." From the ancient "physicians" to contemporary geneticist and exobiological forays, these paths of belief and inquiry have raised the scepter of many an inquisitor, censor or sovereign. They have also received the blessings of presiding clerisies, whether that of the Holy Roman Empire, the Vatican, the US Bioethics Commission, DARPA, or the funding appraisers at Monsanto and the NIH. In the secular context arising with the emergent natural sciences of early European modernism, the debate over plenitude dared to question the nature of the material universe in general.

As illustrated by the intense interests at stake in the renaissance *Lynx* Society's dissection of the two-headed calf, there is a continuing cultural and research fascination with the potential or plenitude of a universe which may exhibit unfathomable varieties, both qualitative and quantitative, in any presence of forms of life. Often extending anthropomorphic images, this desire to capture and to calculate has also sought after presences of life that mirror and imitate it in "intelligent" embodiments.

The enduring origins-of-life debate, including the question of "fitness" and various articulations of the anthropic principle, is an anxious expression of the debate over the nature of life itself. The genomic paradigm discussed in the previous chapter emerges within a generalized social instability and uncertainty about the integrity, viability, and mutability of living systems, ecologically situated in systems of sustainability confronted by human industrial and technical intrusions. Public interest, investment, and participation in genomic and astrobiological research (as in SETI and Galaxy Zoo programs utilizing participating home computational power) may also present responses to the degree to which contemporary cultures consent, ignore, or resist their potential implications.

The perennial fascination with Mars marks a convergence of two formerly distinct scientific modes of inquiry. For many investigators the fourth planet represents what William Sheehan and Stephen O'Meara consider an "untamed intellectual frontier. It is a region untenanted and waiting to be claimed and challenged by our ideas" (2002, p. 12). In NASA's troubled history of explorations and funding new disciplinary borders have been crossed, conjoining a biological and astrophysical will to purportedly disinterested theoretical research. The red planet promises the allure of a unified science not limited to the empirical domain of terrestrial life processes and forms. As Dick notes, the scrutiny of Mars and the search for extraterrestrial life-forms was driven from an epistemic curiosity striving for unified scientific knowledge. In the 1950s the "impressive progress in the study of terrestrial biology did not allow one to separate the contingent from the necessary in living systems, including its chemical basis, the role of proteins and nucleic acids, and probability of life's origins in the first place" (p. 322). Mars became the sign of a "true general biology" that would bring together the emergent knowledges generated by innovative exobiological and astrobiological research programs.

SETI and Lynn Margolis's Serial Endosymbiosis Theory (SET) illustrate the desire of this emergent curiosity. They are theoretical projects that derive from the scrutiny of the place of protists, unicellular organisms that conjoined bacteria within the confines of cell membranes. SET analysis claims evidence for symbiosis as a major dynamic explanatory process largely constitutive in the development and differentiations otherwise explained by evolutionary theory. SET attempts to provide an explanation of evolutionary processes in terms of cohabitation, incorporation, and ingestion of bacterial organisms into cell structures, formative of mitochondria, chloroplasts, and other cytoplasmic organelles.

Proposals like the SET hypothesis have emerged with an interdisciplinary questioning of the degree and quality of differences between living and non-living systems. Originating in a tradition that pitted mechanists against vitalists, the question of an ontological difference between the living and non-living has constantly stirred dissensus. Where mechanists like Huxley, who proposed a protoplasmic model as life's defining material substrate, defended the reduction of biological complexity to principles of physics and chemistry, neo-vitalists, since Bergson and Driesch and perhaps Wiener, variously invoked energetic systems, evolutionary, autopoietic, or informational, that were never subject to empirical analysis. After Huxley's protoplasmic model was replaced by colloidal biochemistry at the end of the nineteenth century vitalists were on the retreat. The first documentation of the operation of viruses and enzymes established the new biochemistry paradigm replete with its enzyme theory. For reflective practitioners the question of the divide and origins of life remained open. Influential biologists like Pirie thought that definitions would remain elusive and efforts to establish a firm border futile, making the claim that biologists could only disclose "borderline systems." For Soviet biologist Oparin, the gap between the living and the non-living was narrow and porous, with no "fundamental difference" between them. This biological and ontological coincidence was evidenced in basic structures, metabolic processes, reproduction, and responses to stimuli, with none of them inherent solely to living systems. Oparin and Haldane stated unequivocally that life could arise from non-life.

The ancient speculation over plenitude has been rejuvenated in astrobiology. The belief that "wherever life is possible, given time, it should arise," or what Dick terms the "gospel of abundant life" propels NASA's explorations of Mars (p. 348). At the end of 1990s Bruce Jakosky (1998) summarized the accumulated evidence for the presence of water on our neighbor. He claimed at the time that there was compelling data from the Mars Orbiter Camera's images indicating "erosional gullies and depositional fans resembling features produced on Earth from the seepage of water in the form of springs" (p. 13). Regions such as the Nigral Vallis are sinuous channels "thought to have been formed by liquid water" (p. 13). The referent remained earth like geological, thermal and climactic patterns. Anthropomorphic images preside over the incessant search for "Goldilocks's zone of 'habitable' exoplanets." One anomaly that breaks terrestrial equivalents is the apparent distribution of the gullies. Against expectation these formations had not been found at the warmest latitudes, such as the equator, or along terrain sloping toward it, where surface temperatures would most likely accelerate any geological process that would account for liquid water. These gullies appeared rather at middle and high latitudes as well as on slopes flowing toward the Martian poles. He urged continuing exploration for water on present day Mars, following the orthodox association of life-forms with its presence. The images of canyons, impact craters and escarpments strongly suggested its movement in the planet's ancient history.

National and venture corporate funding partnerships at NASA, the European Space Agency determines who and when such projects into space and the question of life will continue. A NASA launched Mars Rover, replacing the Spirit and Opportunity probes, named "Curiosity" is intended to further detect the soil chemistry of the planet. After the cold war that motivated much of the Apollo moon missions, there appears to be dramatically less governmentally driven interest in space exploration. Hubble and its successor, the proposed, James Webb telescope system, have consumed much of the astrobiology budget for years to come. The attempt to project American and capitalist superiority into space motivated early NASA missions. As a review of these missions reminds us, the

> Apollo programme came hot on the heels of the duck-and-cover, getting underway just before the Cuban Missile Crisis. The American fear of death rattling down from the open skies gave these en a purpose as saviours, as the people who could show that American rockets could go further than anybody else's (O'Hagan, 2010, pp. 13–14).

It was never a purely blue-sky pursuit. In today's money the entire cost of the "Apollo programme was spent in 540 days of the war in Iraq. Perhaps it was an unexpected result of our lunar wanderings: in the end they marked the limits of the old capacity for wonder, showing, humanity that it was, at base, rather provincial. Craig Nelson considers that "perhaps the most important reason for going to the Moon, was that the space race kept the Cold War cold" (in O'Hagan, p. 13). Entrepreneurs, especially aerospace contractors like Martin Marietta and Lockheed, secure their role in the militarization of space, taking the initiative given them, opening the universe to privatization.

To "follow the water" was the mantra and starting point for exploration of the planets. Initial readings from the Galileo confirmed predictions of abundant water in the form of a liquid ocean on Jupiter's moon, Europa. Beginning with the Ames exobiology program the first Viking expeditions were charged with the task of aiding the search for the origins of life. Intriguing evidence suggestive of processes conducive to the formation of known forms of life were and continue to provide some empirical justification for the massive, often disastrous, projections into frontiers in space like those presented by the Challenger and one of the space shuttles. Among these are the implications made from prebiotic materials, including analysis of meteorites, comets, dust, and the atmosphere of Jupiter and its moons (including Titan). The much investigated Murchison meteorite in particular sparked considerable curiosity. As a very rare specimen, this meteorite is identified as a carbonaceous chondrite, which in this instance was identified as containing 74 amino acids, 55 of which do not exist in any living terrestrial organism. In addition, these amino acids are structurally different from the earthly counterparts, their molecules often branching right handed from an axis in contrast to the universal left branching "chirality" of terrestrial forms. An uncovered carbonaceous chondrite that fell near Tagish Lake in Canada's Yukon also promised to offer supplementary evidence of amino acid and prebiotic material, including "pre-solar nebula" material that antedates the formation of the solar system.

Explorations like the European Space Agency's "Mars Express" launch in 2003 were intended to continue the line of inquiry first begun with the Viking landings. Everett Gibson, at NASA's Johnson Space Center in Houston, is one of the discoverers of the controversial Martian meteorite ALH 84001 that drew widespread, but poorly supported evidence for microbial traces from the planet. Involved in continuing projects of exploration of the solar system, Gibson is convinced that the Mars Express landing would be able to determine conclusively whether life-forms have ever existed on Mars. The Beagle landing unit contained a methane detector, seeking to read for the "signature" of this gas, as evidence for subsurface biota.

Projects like those attempting to disclose the nature of life-forms in general, or as illustrated by proposals to "terraform" the Martian surface through nanotechnological "replicators," are instances of a calculative and instrumental curiosity. The red planet may serve well under the rule of an alien King Midas. It is an acquisitiveness and impulse to assemble diverse objects, whether material or conceptual, into a supervening coherence. In the peculiar history of space flight and research, this passionate pursuit to exceed limits is explicit. The levels of scale of the technosciences now prize the infinitesimal or the infinite, from strange and "charm" particles and Gell-Mann's quarks, the fecund dimension of nanotechnology, to cell line interventions in genomics and proteomics, and the optical spectra devices seeking earthlike "exoplanets." All potentially interfacing levels of scale are rendered "readable" by imaging instruments. A vast array of "visualizing" technics opens dimensions for the extension of thought. Frontiers and barriers are being crossed, causing mixtures of euphoria, resistance and anxiety in their wake.

The significance and consequences of research in fields like genomics and nanotechnology are not yet discernible or decided. As imprints of the contest over their significance and incalculable potentials (as in the theologically inflected controversies over stem cell research and development), costs and profits for some, the agenda of contemporary inquiry is beset within rivalry and contention. These practices are a patchwork texture that in many ways continues the projects of European empirical science's "insatiable curiosity" and empire building (Nowotny, 2008).[6] The modes of knowing, like their epistemic objects, remain impressed by globalizing economic systems that persist in the commodity reproduction of sameness. The promise and threat of a world made increasingly unified permeates the public reception of technologies like genetic engineering. At the same time research and development, particularly in high-tech, IT fields, is increasingly outsourced, or, in corporate rhetoric, "rebalanced," within developing economies that have stressed the growth of knowledge work, with a pedagogical determination to emphasize the teaching of mathematics and sciences. It is a world still unable to structure an equitable distribution of basic resources, including food, health care, water, and shelter. The new technological capacities must be considered in the context of access to both necessary and desirable resources, and their implications for environmental change, at the global, regional, and individual bodily levels of experience.

All these projects catalyze potentials of time, space, "minds," and embodiment. They wander beyond sensible, phenomenal access. Empirical documentation of cosmological and biochemical processes must harness a realism still derived from the grounding principles of a rationalist, visualist scientific culture in the ordering, industrial production and organization of physical and living processes. They all share (with perhaps the exception of the important strangeness of quantum mechanics) what are purported to be grounding laws and processes at an infinitely expansive and infinitesimal scale. Even should much of their hyperbolic and premature claims come to be verified, their significance as explanatory schema remains undetermined. From my phenomenologically accented critique again, they are likely to remain mechanistic, abstracted from the sensory lifeworld's contingent dynamics. The technosciences appear relentless in their commitment to disclosing symmetries, patterns and designs. A technological will to power expands essentially unfettered from any sociocultural and embodied moorings; each marking a resistance to limits in what constitutes a denial and challenge to finitude.

The privilege of reflecting on knowledge is an attribute of specific modes of knowledge production. Disciplinary practices such as genomics, bioengineering, artificial life, and exobiological research programs, granted institutional prestige and funding, generate activities that themselves become self-reproducing, differentiating networks. Theoretical curiosity remains in all instances structured by intellectual labor within academic capitalism's particular disciplinary divides and alliances. The sustained modes of critical inquiry that I champion yield their theoretical energies into affirmatively partisan communities of practice. The spectrum of intellectual labor that has been, at least in the United States since the 1990s, regarded as theoretical desire, has taken on intensely politicized interest as well as speculative and instrumental personae. These are evident in contemporary research agendas as they range across and combine utilitarian, Promethean, socially meliorative, or "blue sky" speculative projects. The first are evident in the rush to find drug development and markets deriving from genomics research, the second evident in therapeutic and clinical access for treatable health and environmental conditions and issues, and the third illustrated in the search for a unifying physical principle in the Large Hadron particle collider project seeking the elusive Higgs boson. These communities of practice and intellectual labor are rationalized, designed for target market utility, (though theoretical physics and cosmology remain valorized), functioning within networks and patterns of research and technical investment. As Rajan's ethnography well documents, this process is standard practice for thousands of small start up "upstream" genomic companies actually conducting the fundamental laboratory investigations later taken "downstream" by the multinational pharmaceutical companies.

In both the molecular biological and exobiological research programs the question of the limits and place of the human have moved center. Both evince a desire to determine the conditions, units, and structures of life-forms. The molecular vision of life incorporated what Keller (1995) and others call the "central dogma" of DNA as the transmitter and vehicle of all pertinent information concerning the possibilities and expressions of life (p. 89). The genetic model, as developed from the earlier molecular biology programs of this century, have generally exorcised embryological research projects, giving pride of place to complexity in computationally based investigations of protein formation, the nucleus of cells and to the chromosomal architectures of organisms in funding, publicity and prestige.

Utility and research are also coupled in the intellectual property and patenting structures to which universities in America have enthusiastically engaged in since the 1980s as effects of the Bayh Act. In its uses and "useless" projects alike, theoretical curiosity assists, in the neocolonial and globalized production of knowledge's surplus value. The trope of a jellyfish Medusan internet, or of digital "network" cultures, prevails. With the emergence of these alliances of research and development the ravening of capital increasingly incorporates academic and intellectual labor in the sciences at the expense of dwindling commitment to the liberal arts and humanities. There is a vigorous integration of institutions that aligns multinational agricultural, drug and pharmaceutical corporate boards, as in the case of the partial supervision of doctoral theses by Novartis in a graduate program at the University of California, Berkeley. Entrepreneurial "economic empowerment zones" newly refigure school districts. These personnel and practices entangle relations of knowledge production, academic—military partnerships and the entrapment of a socially valuable research. Conservative and reactionary resistance to the very viability of the public sector intensifies the tensed ambiguity between the natural and the artful, the constructed and the necessary.

Challenging limits is never really an epistemic, bloodless battle. A critical and material curiosity is embodied in lived conflicts and the problematics of the age. Its finite situation and context is tested and tests, often at considerable risk, the structures of necessity and contingency in social, political and corporeal experience. The ideal of disinterested knowledge sustains the myth of a leisure class, of a contemplative life enabled by the idle pursuit of the purely theoretical and speculative. This myth sets into stone an absolute and objectifying gaze. It remains a potent and disablingly self-serving mythos. It places emphasis on *theoria* and *episteme*, in hierarchical prominence to a genuine *techne* and Orphic crafting in the media, tools and ways of knowing by which we can make and remake culture. The Promethean will of the technosciences and our current instrumental episteme is not inherent to the phenomena, content, or methods that its practitioners utilize and pursue. It need not endlessly eclipse the sensible realm and of its artisanal, aesthetic, and manual counterparts. These are not and never can be regarded as subordinate domains of our knowing and cunning engagements with our living environments.

The place of technics, discussed in Chapter 6, can assist in reappropriating existing partitions of knowledge and practice. Regarding the technoscientific discourses and practices of bioengineering, genomics, artificial life, and exobiology and extraterrestrial studies, I wish to attend to some of the contemporary resonances of theoretical curiosity and its contested place in social relations, the public sphere, and the powers of thought in and about the nature of life.

Does Technology Think?

In the influential Heideggerian (1977) critique of technology, science has lost its bearings and relation to open being. It is a faint echo and expression of the potential of making and becoming in the material world. Technics realizes creative acts of thought in any of the arts of life. It performs "calculating" and instrumental practices without ontological bearings or concern for the bounty and heterogeneity of living beings in specific historical places and times. Science becomes "research"; in its working regulation through methodical order and procedures, scientific inquiry is reduced in its ability to scan the world it would seek to know and explain. Inductive and deductive methods alike, primary and careful generators of technoscientific processes, persist in an unregarded mapping of worlds of inquiry. It is a mapping or "picturing" made in its own image. (Heidegger often asserted an auditory "calling" in contrast to the long tradition in which visuality has dominated.) The value of thinking along the lines of this peculiar critique is that it brings into question the pursuits of technological desire. Is it limitless or bounded? What drives an apparently inexhaustible reserve of the power to generate new productive forces, potentials and ways of remaking the texture of experience?

In modernity, at least since Bacon, Descartes, and Suarez, formal ontologies and objectifying, rigorous methodologies gained ascendancy. Since its early modern inception scientific curiosity has been highly structured in visualist representational models. As in Sohn-Rethel's (1977) analysis, it is structured along mathematical or axiomatic grids. Homogenizing and non-sensible mappings of phenomena and processes regulate work in the natural and human sciences. In the analyses Heidegger provides, science now seeks to make its own processes and interventions secure. This is made possible by a traditionally overarching compliance with contentless methods that grant a high degree of autonomy, sense of realist transparency, and auto-referentiality to scientific communities' rules and maps. Its representational schema, models and propositional or deductive methodologies are constructive and temporarily functional. These representational models, maps, or "model organisms" (for genetic research) factor within their own subcultural systems, and constitute, in effect, rules for a surveying scientific practitioner" disengagement from the phenomena under "inspection." The formal mapping is too often unreflected unless dominant models are challenged and their research programs reoriented when necessitated by countervailing empirical findings, new maps and pictures. These pictures are generators of research findings themselves, augmented by technical optics, instruments and devices, from microscopes, "spider DNA nanobots," space probes like the Hubble or European Space Agency's Planck orbiters, bubble chambers to centrifuges, turbines to fMRI imaging technologies.

Heidegger, speaking his own idiom, asserts that a "standing reserve" exists in the modern world of technological dominance. He is referring to a technological desire, as a potentiality that remains suspended, "reserved" and kept from coming into actuality. It is the power of potentiality itself. A virtual "reserve" provides the ontological support for the vast and pervasive technical regime. This is an era of hybrids of culture and technics. In his critique it is one shaped by a basic amnesia, a forgetting of origins. From his perspective the technological itself has no core, essence or meaning in itself. Scientific inquiry itself, reduced to research, does not think its own foundations and purposes. The technological comes to reduce life to an ordered and calculated, instrumental mode of existence, shaping cultural memory and the power to think about our existential participation. A self-perpetuating technology now indicates a specific historical symptomatic turn for an originary, creative techne. It is potentiality held, never fully realized, operating outside any accountability or concern for its powers to affect the material and experiential actuality of everyday life. Merleau-Ponty (1997) shared this view, writing that

> science manipulates things, and gives up living in them. Operating within its own realm, it makes its constructs of things; operating upon those indices or variables to effect whatever transformations are permitted by their definition, it comes face to face with the real world

> only at rare intervals. It is, and always has been, that admirably active, ingenious, and bold way of thinking whose fundamental bias is to treat everything as though it were an object-in-general, as though it meant nothing to us and yet was predestined for our ingenious schemes (p. 121).

This regime of representation erases or "applies" an original theoretical "seeing."

For Gadamer, Heidegger, and McNeill, theoretical practice, resounding with its rich semantic and classical overtones, comprises a broad and intrinsic spectrum of human activity. It signified, variously, a seeing, spectating at an event, observing, and serving as an envoy, like Herodotus, to foreign lands and peoples, and as a participant in the practices, whether sport, art, or epistemologic dialogues, like those with Socrates in the *Meno* or *Theaetetus*. Technocultural production obscures its own origins as a practice in and on the world. It can work its transformative potential in salutary ways that need to be acknowledged and recognized as contributions to human and animal preservation. When it becomes effectively autonomous and self-regulating, or organized solely on the basis of the interests of private enterprise, the link to the ecologies of experience and material reality are severed.

Representations give shape to practice through their formation of foundational "projections." These projections usually remain obscured or forgotten by the methods that, especially in traditional natural science, they drive through their sedimented schema, models, propositional, or deductive and axiomatic logics and laws. As prisms, projections set the refracting vision of science and technology. Theory, historically situated, loses sight of its own origins, foundations, and potentials. As a historical desire, technology appears outside of human condition, more of a determinate force in the production of things and techniques than in giving meaning or aesthetic pleasure and sense to everyday life.

Technology appears to occupy its own temporality. Disguised as an ahistorical capacity to produce limitlessly, it gives expression to a desire to capture phenomena in its own image. It assists in the overproduction of technical commodities and instruments. In the visual representation of its "capture," it produces research optics, the media advancing quantitative changes in scale of observable phenomena that also construct a surplus of means for generating or discerning new epistemic objects. It is a desire that does not acknowledge or return to the activity or springs of its own practices in human uses of the nature, the making and reworking of what occurs naturally, as in artisanal techne. It conceals, for Heidegger, the concern, care and *curiositas* that might encourage us to participate in reciprocal relations with natural, animal and intersubjective environments. It conceals, through the primacy of method, an ontological drift, a vertigo of productivity, from experience, reducing it into abstraction and virtuality. Theory comes always to be applied to praxis. Basic research is instrumentalized. Human creative working and forming is displaced as only direct, calculatingly productive involvement with things becomes dominant. Techne, in its originally capacious sense of the sense of seeing both the visible and the invisible, of the material and the ideal in things and phenomena, is lost. In reducing theory to application, to use values and profit, it is a perpetual "surplus" of forms that control, order or deny their connections to life and its enhancement.

As a complex of changing practices, programs of scientific research are sustained and developed by an abiding theoretical curiosity within its varied communities of practice. The work within specialized fields of inquiry, as found in primatology, neuroscience or particle physics, for instance, proceed with the constant vigilant ability to review and critique findings, data and making decisions about what might be fruitful fields to pursue. Their own methods and instruments though do not always come into question. Though it is necessary to question the process of accumulating evidence, findings and "raw data," reflection upon the directions and purposes of research programs and their implementation should occur as an integral component of any qualitatively interested project. Scientific research, especially under the spell of a disembodying technological productivity at play across a networked globe, is at once a way of seeing the world as well as seizing it in contingent representations.

Modern technosciences are extensions of a pervasive and invasive "inspection" of a world it would capture for a future that never comes. It is also a desire that compels the ancient's *noos*, the cognitive and intellectual capacities, into an aggressive form of contemplation. Playing with its etymology, contemplation means here a cutting up or analysis of the phenomena of existence and inquiry. As a cutting up, it reduces this primordial seeing and knowing. It is a concerted and organized aggression, an "attack" against the natural becoming, the physis, of things and our place. For Heidegger (2002), this is a grasping and a fundamental breaking up of phenomenal experience into units of manageable analysis. It is an ordering and regulation that levels and distances scientific inquiry from the world it would seek to explore. The "uncanny" thing about the sharpened lens of this contemplative aggression, or drive, is that it never stops. It is an integral part of an overarching panoptic gaze on all things and processes—a survey of the entire "world picture." This grasping also seizes time. While "mere" curiosity is contrasted by Heidegger with a pure seeing into the essences within appearances, of how things come to appear, technology appropriates finitude and time. Curiosity is always a preliminary and, at best, a partial and passing opening to phenomena. It does not concern itself with anything, seeking only the new and actualities. Modern technology, however, is regarded as engaged and concerned with material and phenomenal actuality as far it is made secure, produced by its own "will to will." It maintains actuality in the realm of an ever suspended possibility and virtuality. It is virtual in that technological ordering, instanced by algorithmic and digital programming, is one in which substitution, iterability, replacement or commutability are operative. It is a calculating rationality that is invested in its own potential to produce. Human sensuous and perceptual experience and the knowledges that might emerge from its treasures are encoded in the machine languages of ones and zeros, one in which we live as digital subjects.

Technoculture is an alternative framing and instituting of the theoretical life advocated by Aristotle. In the formulation of his claim that the theoretical life is the most excellent and desirable praxis, Aristotle did not sever its place from the practice and purposes of social and political life. The life of theory and the human expression of its pursuit, *zoos theoretikos*, was not in opposition to the necessary and desirable practical uses of knowledge or phronesis in the life of the citizen, the *zoos politicos*. They form a logic of limits that is not polarized. The theoretical attitude of the classical episteme embodied in Presocratics like Thales, who stood in ridicule for his disregard for the everyday, would stand in wonder at the appearance of things. For Aristotle the life of theory, a luxury for an elite who did not need to worry about their material needs, was also conducive to the fulfillment of the praxis of friendship, of lives lived among and with others.

Methods provide primary projections, mappings or "pictures" of the phenomena they objectify. Epistemic objects are historically located, defined and made into resources for basic research and for their potential uses and management. Heidegger's trenchant critique is certainly limited itself. His strategy is to work to unframed, to destruct representational methods and practices. He asserted that this representational regime is unique to modernity. This is especially pertinent to the early scientific revolution. The experimental methods of Baconian science, including heat, magnetism and chemistry, were certainly premised on mathematical models. It is the split world, one in which a sovereign subject has come to claim a knowledge of a world outside of itself. Nature is to be tamed in large part by our "pictures" of it, in a mastering pursuit of timeless and placeless truth. This pictured world hides its own sources. It conceals the actual praxis, at once theoretical, mechanical, epistemic and manual, that have formed it historically. The Subject, armed as the transcendent cogito, exists entirely apart, a disembodied scion of the Soul; within a representational, visualist, and formal ontology, at least since the sixteenth century, it is formed within structures supporting the phantom of an episteme of calculation and abstraction.

Rouse (2005) usefully challenges the pertinence of Heidegger's critique in light of the phenomena of interest in current sciences. While there is abundant evidence of an uncontrolled theoretical and speculative curiosity in which the "domain of experimental manipulation and theoretical modeling seems to expand without apparent bounds," Rouse suggests that most "truths about the natural world are of no scientific significance whatsoever" (p. 18). Technosciences attend in large measure to specific phenomena, theories, concepts and models, like the interest generated by the "November Revolution" in physics in which "weak" neutral currents were observed, or the studies concerned with the typology of cancer cells' structure and function. This is also illustrated in the crossing of the boundaries marked by postmodern quantum mechanics, in that it "forsakes mathematical elegance and systematicity in taking advantage of multiple formally inconsistent models simultaneously to encourage a physics of irreducible complexity." For Rouse, postmodern quantum mechanics rejects the quasi-theological fundamentalism governing much of recent high-energy physics, abandoning the quest for a unified "Theory of Everything" in favor of more local, situated comprehension. Similarly, the phoenix-like emergence of developmental biology from the ashes of embryology, and concomitant eclipse of genetics by genomics, challenge the now-familiar conception of genes and DNA as the calculatively controllable "secret of life and biological surrogate for the soul" (p. 19).

Missing from Heidegger's influential critique of technology is an appreciation of the inseparable reciprocity between technics and cultural production. The artifices of techne, of the inanimate and non-living, are essential supplements to cultural memory and continuity. Heidegger retreated from the visible as an entrance to reinventing everyday ontic being. As I described in Chapter 4, the everyday provides the resources and traces for a critical curiosity that would transform social life. Critical curiosity of the everyday must assume that now technics is the convergent site across pedagogies and cultural fields. Material practices are subject to constant renovation and alteration generated by technical systems. The tools and instruments cultural production actualizes are differing ways of acting in and within the ecology of lifeworlds. There is always a responsive relation between technics and cultural continuity. This is not, however, to say that the technical determines material culture and the limits or forms in which knowledge is practiced. Sociocultural changes are not expressions of any determinate telos.

While Heidegger's project was intended to interrupt the blind pursuits of a reductive "research," insights like Rouse's offer a perspective more attuned to the complexity of the phenomena studied within the technosciences and the variety of research communities and interests. Rouse's examples are also incalculable instances of the "coming to presence" of the phenomenal world of perception that so occupied Merleau-Ponty. For Rouse, "such cases cannot be appropriately regarded as impositions of a predetermined orientation toward calculative control upon nature as a plastic resource, for what it matters to understand calculatively .... Such shifts instead reflect an openness within science to allowing things to show themselves intelligibly in new ways" (p. 19).

Scientific practice is a dominant twist in the historical fortunes of a techne inherent to all human work and cultural productivity. In Heidegger, science, whose origins might begin in the pursuit of knowledge of processes and structures in nature, including the human sphere, separates itself from this purpose in a "will to will." The desire to know becomes applied, utilitarian, forced into material production and reproduction. In this technologizing of a desire innate to human experience, it loses the "sense" that phenomenology seeks to recover. Since Husserl (1970) as discussed earlier, science has come to signify a general crisis within the modern world. It has lost its sense of the meaning of its work and projects, of the "immediacy of its own praxis in the face of the world" (p. 40). It is the historical signature of the modern era in its representational power to order, control and plan. Technology, and the technics at play in the globalized economy that it supports, give contour to the present and in large measure determines the future. Technology mutates ontological and theoretical sight. It has abandoned the font of wonderment, the gaze into the invisible within and beside the visible. As a "cutting up" this form of contemplation is an analytic tearing apart of the unity and presence of the multiplicity and emergent actuality of material existence. It is a rent in the fabric of the experience of the "originary look" and of sensuous being and becoming.

In terms of the pursuit for the limits of both visible and the invisible it is a contest whose empirical particulars are avidly pursued with a discerning precision for disclosing foundational frames of reference. These unifying symmetrical forms can be signified by combinatory relations and transformations between matter, energy or information, data or code. The search is unrelenting, and like the nested spheres in the cosmos of Plato's *Timaeus*, perhaps endlessly recursive. A final, fixable structure or process, integrating function and form, structure and energy, mechanism and flux, is pursued in determining universal forms and structures. They find rehabilitation within new guises and rhetorics, such as the new interest in complexity and emergent systems. The figures of chaos and complexity theories have been reductively conflated and appropriated for mass consumption.[7]

Limit Forms

The extent to which we can make knowledge part of our cohabitation, in ventures that either contain or surpass the limits of our finite embodiment, remain open questions. They are not idle, speculative questions. How they are answered or left unresolved will give shape to nothing less than the future of humanity, of all living things and the earth itself.

The desire of a theoretical search for the ultimate roots and foundations of biological and physical processes remains a driving force across research programs, disciplines and pedagogies. At both the cosmological infinite and the "infinity of the infinitesimal" there is a Promethean will to power in giving form to the formless. They are inquiries with long historical, theological, and metaphysical imprints. It is a complexly accented historical pursuit in which form itself is in question. The historical circuits, trials, and discoveries of theoretical curiosity provide an ambiguous legacy for the pursuits of knowledge. Participating deeply in our cultural imaginations, these immensely varied and singular pursuits I identify as our contemporary variants of technics—and the constraints placed upon them—bear the imprint of a reserve and potential that may never be fully materialized. The desire for knowledge is a potentiality that perhaps will always exceed the limits of any social imaginary, the patterns of everyday life and the conflicts between tradition and innovation.

In response to the global financial crisis in the first decade of this century, Richard Seaford (2009) argues for a new "culture of the limit." Capital flows, always mercurial, surge and dive in continual crises. Seabrook observes a generalized refusal of any limits. This is evident in the deregulation of markets, financial bailouts, corporate bonuses, as well as in a general cultural homogenization of space and place, as illustrated by urban planning, hotel and restaurant chains. He contrasts classical Greek culture's attention and anxiety regarding the introduction of money with our own globalized economies, drawing out a parallel between the accumulation of wealth and knowledge. Contemporary cultural, scientific, and economic life are "characterized by a hostility to closure (limit) in various spheres, economic, metaphysical, conceptual and narrative" (p. 14). Theoretical projects, across the disciplines, also resist closure. In the classical world of the Platonic dialogues, and in Aristotle and Aeschylus, there was an expressed concern for the historical experience of the change from an exchange economy to its monetization by granting priority to normative limits. Plato's *Philebus* states that limits must control the unlimited. Aristotle, whose cosmology was comprised of a closed universe, declares in the *Nicomachean Ethics* that the "good" can only be found in the limit. These are he historical boundary markers and the original meaning of method; resources for thinking, along with Heidegger, the place of science and technology. The notion of a culture of the limit is timely. For Seabrook, the Greek culture of the limit provides a place that allows us to "see the oddness, the historical contingency of the *lethally limiting unlimitedness* of our own economic, social practices and theory" (p. 15).

Odysseus personified the surpassing of limits. For Renaissance explorations and inaugurating the Enlightenment discussed in Chapter 2, Prometheus and the wandering seafarer have been emblematic. The resonances of the myth of Oedipus also play a significant, though enigmatic role, in the preoccupations of inquiry and discovery. As the mythic figure of one who lives and sees with "one eye too many," he embodies a legacy of ambiguity in confronting irresolvable contradictions. Reappearing in Kant, Oedipus reveals the contradictions to which human reason is exposed when it goes beyond its own limits, goes beyond the limits of finite experience, and attempts to "take up a position beyond the sensible world" (in Lacoue-Labarthe, 2003, p. 17).

For Lacoue-Labarthe, his is the eye that discerns a space within which to think the tensions of contradiction. Oedipal knowledge is at once the knowledge that animates both desire and labor. It is the undercurrent and momentum for theoretical curiosity in all its lived embodiments, its confrontations with limits and its often tragic encounters with contradiction. In Lacoue-Labarthe's suggestive reading, Oedipal knowledge is a figure for the unrepresentable. It is an unbounded but substantive originary energy. It propels techne. It is a kinetic spur for the corpus of the metaphysical tradition of the west. It is the groundless ground for all human practice. Echoing Heidegger's critique with a twist, he asserts that "modern technology is the oedipal realization of the metaphysical" (p. 17).

Forms are being reembodied virtualized. Representation has met its limits. In question and subject to intervention are the forms of bodies, concepts and representations of the human and the animal lifeworlds. These disputed regions of sense are all potentials for acquisitions and capture, in Heidegger's sense. They underlie the contested and politicized criteria for inclusion in any cultural canon. They are put into practice in the methodological and technical protocols, codes and programs by which we are oriented to reading the world. They provide the usually unreflected coordination of the "natural attitude," how we are inscribed within relations to or against nature, myth, and law. Forms are no more transparent or innocently given and practiced than the historical conditions that make them possible. The intense detection of the mobility of forms, structures and aspects of energy in the now dominant technosciences may refract the interests generating their projects; the volatility and mobility of forms of matter and life, their very indistinction, may be traced as an analogue to the global shifts, reserves, and starts of financial capital. Paul Rabinow (1999), astutely commenting on genetic research, observed that

> from time to time, and always in time, new forms emerge that catalyze previously existing actors, things, temporalities, or spatialities into a new mode of existence, a new assemblage, one that makes things work in a different manner and produces and instantiates new capacities. A form/event makes many other things more or less suddenly conceivable (p. 180).

While a critique of social theory and its production is implicit, this project is not intended primarily to advance another variant of critical theory. Though I have stressed research and inquiry with a phenomenological accent, this essay is an open invitation to readers and researchers to consider the origins of theoretical practices in their sources of social conditions of possibility, in their historical contingency, consequences, and potentials. I only wish to affirm the importance of theoretical curiosity as a constant critical engagement with the direction, purpose, and limits of any research program. It is a name for the struggle of reason against dogmatism, superstition and the "commonsense" that uncritically accepts the given as timeless. Theoretical curiosity, as Blumenberg documented, is also a name for the practices by which science as knowledge came to emerge and by which we recognize our "modernity." It is a goad and spur that recommends an open pedagogy for researchers, teachers and students in all disciplines. I have worked this theme from a historically informed and phenomenological perspective in order to suggest that the texture and qualities of our experience are our responsibility, creatures of our creative practices. Theoretical inquiry, in its materiality, is embodied thought and activity, refusing to remain speculative and idle. Like Marx we have the historical conditions given to us and we are partisan to the making of the future, not just its interpretation. It is not the answer, concept or sign for anything more than the ways of knowing and questioning that allow for choices in composing the texture of our world. Curiosity can take us anywhere or nowhere; it is an openness to the multiplicity of intersecting futures that appear on the horizon of human projects.

I want to conclude in affirming the material premise of my advocacy for an aesthetic and phenomenally informed appreciation of theoretical inquiry. It is a catalytic element of intentionality; it makes forms in the situated ecology of thought's embodiments. As responses and solutions to evolutionary conditions, including environmental changes, life-forms are continually subject to new perturbations, challenging homeostasis, conceptual order, representation, and equilibrium. Self-reproducing and adaptive, biological embodiment requires continuity and coherence within divergent and changing environmental influences. With respect to these environmental and evolutionary factors, I regard curiosity as an aspect of species being, constantly appearing in phenotypes becoming different. Bodies take form in unstable stabilities and adaptive niches, at once epigenetic and technical. Curiosity is generated in the same layers of specific, geographical, historical, and technical milieu as are languages, images, and affect. Sense, images, and ideas adapt and detach through the niches of their living milieu; they can pass beyond them but not without them. Out of the places of everyday life and relation, novel forms of experience, thought, and theory can conjoin and take flight.

Theoretical curiosity, driven by critique of everyday life, is a channel between possible and actual modes of practice. A critical retrieval of the abstracting optic and mathematical models, of humanity's most speculative and speculating pursuits, is necessary. "Idle," speculative thought or methods should neither be censored nor made idols of the tribe. While technics are ligaments and tissues of all cultures, an anthropological principle, they are not autonomous and determinative of their forms, processes, and expression. In Gilles Simondon's (1992) terms, technics and material culture function in a "transductive" relation with the actors and agents of a social formation. They converge and convey how memory and practice are primarily constituted. They do not function without resistance, adaptation, and innovation.

The technical is not a contemporary Olympian demon or drunken god compelling mortals to perform tricks. As Stiegler reminds us, Prometheus brought the techne of fire to "unequipped" humans because his brother Epimetheus forgot to grant the human species a distinctive power or trait to survive and flourish among other creatures. How, since the Paleolithic, we set and shape our fires and sharpen stones is a legend of persistence and invention. Any habitable future will take form in the play of shadows and heat sparked by this crafting animal.

Notes

1. See for instance, Pickering, A. (1995). *The mangle of practice, time, agency and science.* Cambridge: Cambridge University Press.
2. See, for instance discussion of this issue in Luhman, N. (1995). *Social systems*. Palo Alto, CA: Stanford University Press; Wolfe, C. (1998). In search of post-humanist theory—the "second order" cybernetics of Maturana and Varela. *Cultural Critique, 30*(Spring), 33–70.
3. Katherine Hayles (1998) makes a useful distinction between "inscription" patterns from "incorporations" or instantiations in matter, whether organic, inorganic, or in compositions of both.
4. Keith Ansell Pearson discusses the connections between capitalism and codes in his discussion of Guattari and Deleuze's collaborative work, especially their *Anti-Oedipus* and *a Thousand Plateaus.* Responding to germane criticisms that their work "disavows" an embodied, human subject, Ansell Pearson responds in a defense of Deleuze's project pertinent to this discussion that it was Marx who proffered a guiding insight in understanding that it is "capital itself which develops as a metaphysics of energy." The translation of Marx he offers from the Economic and Philosophical Manuscripts of 1844 is, "it [politicaleconomy/private property] develops a cosmopolitan, universal energy which breaks through every limit and every bond and posits itself as the only policy, the only universality, the only limit and the only bond." In Ansell Pearson, K. (1999). *Germinal life, the difference and repetition in Deleuze* (p. 218). London: Routledge.
5. Work conducted at the Joint Laboratory for Attosecond Science in Ottawa produces x-ray radiation sufficient to arrest electron orbits. An attosecond is a billionth of a billionth of a second. In "In Pursuit of the Briefest Beat," *Science News*, March 27, 2010, p. 16.
6. Helga Nowotny is the vice president of the European Research Council.
7. They have been facilely integrated into some fashionable aspects of social science and humanities research, working ironically, as a unifying principle or figure for discontinuity, fragmentation, and flux for a purported universe shorn of any certainty except uncertainty. This notion naturalizes aspects of the everyday experience of alienated individuals and communities within societies of the spectacle. At the same time, conflated notions of these theories give rise and reassurance to a principle, law, and patterning of the indeterminate.

References

Agamben, G. (1993). *The coming community*. Minneapolis, MN: University of Minnesota Press.
Aristotle. (1966). *De Anima 2.2.413b4-5*. In R. McKeon (Ed.), *Basic works of Aristotle*. (p. 557). New York: Random House.
Aristotle. (1998). *Metaphysics*. New York: Penguin. Book Alpha 1, 980a; Alpha 2, 982b, 1078b17.
Aronoff, M. (1962). Freud's conception of the origin of curiosity. *Journal of Psychology*, *54*, 39–45.
Aronowitz, S. (2007). *Against schooling*. Retrieved August 14, 2009, from www.stanleyaronowtiz.org./articles
Bachelard, G. (1938;2002). *The formation of the scientific mind*. Manchester: Clinamen.
Bacon, F. (1996). On beauty. In essays. In J. Spedding (Ed.), *The collected works of Francis Bacon* (pp. 175–191). London: Penguin, Routledge Thoemes.
Badiou, A. (2001). *Ethics, an essay on the understanding of evil*. London: Verso.
Badiou, A. (2009). *Theory of the subject*. London: Continuum.
Badiou, A. (2010). *The communist hypothesis*. London: Continuum.
Balakrishnan, G. (2000). Florence/Moscow. *New Left Review*, *3*, 159.
Balibar, E. (1996). *Structural causality*. In A. Callari, & D. F. Ruccio (Eds.), *Postmodern materialism and the future of Marxist theory* (pp. 109–119). Hanover, NH: Wesleyan University Press.
Barabas, R. (2004). *The Being of the phenomenon, Merleau-Ponty's ontology*. Bloomington, IN: Indiana University Press.
Barabas, R. (2006). *Desire and distance, introduction to a phenomenology of perception*. Palo Alto, CA: Stanford University Press.
Beardsworth, R. (1988). Thinking technicity. *Cultural Value*, *2*(1), 70–86.
Benjamin, W. (1999). *The arcades project*. Cambridge, MA: Harvard University Press.
Berry, C. (1994). *The idea of luxury: A conceptual and historical investigation*. New York: Cambridge University Press.
Blanchot, M. (1993). *The infinite conversation*. Minneapolis, MN: University of Minnesota Press.
Blumenberg, H. (1985). *The legitimacy of the modern age*. Cambridge, MA: MIT Press.
Boas, M. (1962). *The scientific renaissance 1450–1630*. New York: Harper and Row.
Bourdieu, P. (1999). *The weight of the world*. Palo Alto: Stanford University Press.
Bourdin, J. (2000). The Utopian Materialism of Louis Althusser. *Graduate Faculty Philosophy Journal. The Revival of Materialism*, *22*, 1.
Brunschwig, J. (2000). Epistemology. In J. Brunscwic & G. E. R. Lloyd (Eds.), *The Greek Pursuit of knowledge* (pp. 18–39). Cambridge, MA: Belknap Press of Harvard University Press.
Burke, P. (2000). *A social history of knowledge, from Gutenberg to Diderot*. Cambridge, MA: Polity Press.
Carruthers, M. (1998). *The craft of thought: Meditation, rhetoric, and the making of images, 400–1200*. Cambridge: Cambridge University Press.
Casey, E. (1997). *The fate of place, a philosophical history*. Berkeley, CA: University of California.

M. Zuss, *The Practice of Theoretical Curiosity*, Explorations of Educational Purpose 20, DOI 10.1007/978-94-007-2117-3, © Springer Science+Business Media B.V. 2012

Cassirer, E. (1978). *The problem of knowledge, philosophy, science, & history since Hegel*. New Haven: Yale University Press.

Cassirer, E. (2010). *The individual and the cosmos in renaissance philosophy*. Chicago, IL: University of Chicago Press.

Cicero. (1994). *De Finibus Bonorum et Malorum* (vol. xvii, book v, pp. 291–503). Cambridge University Press, XVII.

Collins, F. (2010, September 6). In P. Boyer (Ed.), *The covenant. The New Yorker*, p. 65.

Coole, D. (2000). *Negativity and politics*. New York: Routledge.

Cooper, M. (2008). *Life as surpus, biology and capitalism in the Neoliberal Era* (pp. 103–128). Seattle, WA: University of Washington Press.

Crary, J. (1999). *Suspensions of perception*. Cambridge, MA: MIT Press.

D'Alembert, J. (1975). *Preliminary discourse to the encyclopedia of Diderot*. Chicago: University of Chicago Press.

Daston, L., & Park, K. (1998). *Wonders and the order of nature 1150–1750*. New York: Zone Books.

Debord, G. (1994). *The society of the spectacle*. New York: Zone Books.

de Certeau, M. (1987). The Gaze Nicholas of Cusa. *Diacritics*, *17*, 3.

Deleuze, G. (1983). *Nietzsche and philosophy*. New York: Columbia University Press.

Deleuze, G. (1988a). *Foucault*. Minneapolis, MN: University of Minnesota Press.

Deleuze, G. (1988b). *Spinoza: Practical philosophy*. San Francisco: City Lights.

Deleuze, G. (1990). *Expressionism in philosophy: Spinoza*. New York: Zone Books.

Deleuze, G. (1994). *Difference and repetition*. New York: Columbia University Press.

Deleuze, G., & Guattari, F. (1988). *A thousand plateaus, capitalism and schizophrenia*. Minneapolis, MN: University of Minnesota Press.

Deleuze, G., & Guattari, F. (1994). *What is philosophy?* New York: Columbia University Press.

Derrida, J. (1976). *Of grammatology*. Baltimore: Johns Hopkins Press.

Descartes, R. (1989). *The passions of the soul*. Indianapolis, IN: Hackett.

Detienne, M. (1999). *The masters of truth in Archaic Greece*. New York: Zone Books.

Dewey, J. (1929). *Experience and nature*. LaSalle, IL: Open Court.

Dewey, J. (1997). *How we think*. New York: Dover.

Dick, S. (1996). *The biological universe*. New York: Cambridge University Press.

Diderot, D. J., & d'Alembert, J. (1751–1780). *Encyclopedie, ou Dictionnaire raisonne des sciences, des arts et des métiers* (Vol. 11). Paris: Biriasson, Le Breton, Durand.

Dijksterhuis, E. J. (1961). *The mechanization of the world picture*. Oxford: Oxford University Press.

Dillon, M. C. (1998). *Merleau-Ponty's Ontology*. Dordrecht And Boston: Northwestern University Press And D. Riedle Publishing.

Doyle, R. (1998). Emergent power: vitality and theology in artificial life. In T. Lenoir (Ed.), *Inscribing science, scientific texts and the materiality of communication* (pp. 304–327). Palo Alto, CA: Stanford University Press.

Dyson, F. (1999). *The origins of life*. Oxford: Oxford University Press.

Elias, N. (1974). *The civilizing process*. London: Blackwell.

Elias, J. (1994). *Paulo Freire: Pedagogue of liberalism*. Malabar, FL: Krieger.

Emmeche, C. (1991). *The garden in the machine: The emerging science of artificial life*. Princeton, NJ: Princeton University Press.

Feynman, R. (1999). *The pleasures of finding things out*. New York: Perseus.

Fink, E. (1995). *Sixth Cartesian meditation. Introduction (R. Bruzina)* (pp.vii–xcii). Bloomington, IN: Indiana University Press.

Fisher, P. (1998). *Wonder, the rainbow, and the aesthetics of rare experiences*. Cambridge, MA: Harvard University Press.

Foucault, M. (1970). *The order of things*. New York: Vintage.

Foucault, M. (1977). *Discipline and punish*. New York: Vintage.

Foucault, M. (1990). Diferende la societa. Florence: Ponte alle Grazie. In A. Stoler (Ed.), *Race and the education of desire*. Durham, NC: Duke University Press.

Foucault, M. (1994). *The order of things*. New York: Vintage.

Foucault, M. (1996). The masked philosopher. In S. Lotringer (Ed.), *Foucault live: Collected interviews* (pp. 302–308). New York: Semiotexte.

Foucault, M. (2003). *Society must be defended, lectures at the college de France, 1975–1976*. New York: Picador.

Freire, P. (1985). *The politics of education: Culture, power, and liberation. Westport*. Conn: Bergin and Garvey.

Freire, P. (1998). *Pedagogy of freedom: Ethics, democracy, and civic courage*. Lanham, MD: Rowman and Littlefield.

Freud, S. (1953). *Analysis of a Phobia in a five-year-old boy* (p. 10). London: In Standard Edition of Freud.

Freud, S. (1964). *Leonardo Da Vinci and a memory of his childhood*. New York: Norton.

Fukuyama, F. (2002). *Our posthuman future*. New York: Farrar, Strauss and Giroux.

Gadamer, H. G. (1998). *In praise of theory*. New Haven, CT: Yale University Press.

Gadamer, H. G. (2001). *The beginning of philosophy*. New York: Continuum.

Gatens, M., & Lloyd, G. (1999). *Collective imaginings: Spinoza, past and present*. New York: Routledge.

Gellner, E. (1988). *Plough, sword and book*. Cambridge: Cambridge University Press.

Gibbs, W. (2001, August). Cybernetic Cells. *Scientific American*.

Ginzburg, C. (1976). High and Low: The Theme of Forbidden Knowledge in the Sixteenth and Seventeenth Centuries. *Past and Present*, *73*, 28–41.

Ginzburg, C. (1983). Clues: Morelli, Freud and Sherlock Holmes. In U. Eco & T. Sebeok (Eds.), *The sign of three, Dupin, Holmes, Peirce* (pp. 81–118). Bloomington, IN: Indiana University Press.

Goldin, D. (2000). *Planetary reporter*. April: Vol xx. Pasadena, CA: Planetary Reporter.

Grant, E. (2007). *A history of natural philosophy*. Cambridge University Press.

Greenblatt, S. (1991). *Marvelous possessions: The wonder of the new world*. Chicago: University of Chicago Press.

Hadot, P. (1993). *Plotinus, or the simplicity of vision*. Chicago: University of Chicago Press.

Hadot, P. (2006). *The Veil of Isis, an essay on the history of the idea of nature*. Cambridge, MA: Harvard University Press.

Halbertstram, J., & Livingston, I. (1985). *Posthuman bodies*. Indiana University Press.

Hall, J. (2009). The politics of display. *The Guardian Weekly*, *18*(12), 48.

Hardt, M., & Negri, A. (2004). *Multitude, war and democracy in the age of empire*. New York: Penguin.

Hayles, N. K. (1998). *How we became posthuman, virtual bodies in cybernetics, literature and informatics*. Chicago: University of Chicago Press.

Heidegger, M. (1962). *Being and time*. New York: Harper and Row.

Heidegger, M. (1977). *Basic writings*. New York: Harper Collins.

Heidegger, M. (2002). Anaximander's saying. In *Off the beaten track*. (pp. 242–282). New York: Cambridge University Press.

Heller-Roazen, D. (2007). *The inner touch, archaeology of a sensation*. New York: Zone Books.

Hilts, P. (2001). Scientists are starting to add letters to life's alphabet. *New York Times*, *July 24*, F2.

Hirschman, A. (1977). *The passions and the interests: Political arguments for capitalism before its triumph*. Princeton, NJ: Princeton University Press.

Hobbes, T. (1968). *Leviathan*. New York: Penguin, (1.6).

Hofstadter, R. (1966). *Anti-intellectualism in American life*. New York: Random House.

Hugh of St. Victor. (1961). *Didascalicon: A medieval guide to the arts*. New York: Columbia University Press.

Hume, D. (1965). *A treatise on human nature*. Oxford: Oxford University Press.

Hundert, E. J. (1997). The ideological context of Mandeville's rise to fame. In E. J. Hundert (Ed.), *Introduction* (pp. x–ix). Indianapolis, IN: Hackett.

Husserl, E. (1970). *The crisis of European sciences*. Bloomington, IN: Northwestern University Press.

Ihde, D. (2007). *Listening and voice: Phenomenologies of sound*. Albany, NY: State University of New York.

Jacob, F. (1973). *The logic of life, a history of heredity*. Princeton: Princeton University Press.

Jakosky, B. (1998). *The search for life on other planets*. New York: Cambridge University Press.

James, W. (1978). *Pragmatism*. Cambridge, MA: Harvard University Press.

James, W. (1979). *The will to believe*. Cambridge, MA: Harvard University Press.

James, W. (1983). *Principles of psychology*. Cambridge, MA: Harvard University Press.

Janicaud, D. (1994). *Powers of the rational; science, technology and the future of thought*. Bloomington, IN: Indiana University Press.

Jappe, A. (1999). *Guy Debord*. Berkeley, CA: University of California.

Johnson, M. (2007). *The meaning of the body, aesthetics of human understanding*. Chicago: University of Chicago Press.

Joyce, J. (1990). *Ulysses*. New York: Vintage International.

Kahn, C. (1994). *Anaximander and the origins of Greek cosmology*. Indianapolis, IN: Hackett.

Kakar, S. (1976). Curiosity in children and adults: A review essay. *Indian Journal of Psychology*, 51.

Katzenellenbogen, A. (1939). *Allegories of the virtues and vices in medieval art*. Toronto, ON: University of Toronto Press.

Kay, L. (1993). *The molecular vision of life: Caltech, the Rockefeller foundation, and the rise of the new biology*. New York: Oxford University Press.

Keller, E. F. (1995). *Refiguring life, metaphors of twentieth century biology*. New York: Columbia University Press.

Kenny, N. (1998). *Curiosity in early modern Europe*. Wiesbaden: Harrassowitz.

Khayati, M. (2002). Captive words (Preface to a situationist dictionary). In T. McDonough (Ed.), *Guy Debord and the situationist international* (pp. 173–180). Boston: MIT Press.

Kincheloe, J. (2008). *Knowledge and critical pedagogy*. New York: Springer.

Knabb, K. (1981). Notes editorials: L'Operation contre-situationiste. In K. Knabb (Ed.), *Situationist international anthology* (pp. 108–114). Berkeley, CA: Bureau Of Public Secrets.

Knorr, W. (2000). Mathematics; Proclus. In J. Brunschwig & G. E. R. Lloyd (Eds.), *The Greek Pursuit of knowledge* (pp. 235–262). Cambridge, MA: Belknap Press of Harvard University Press.

Kuhn, R. (1976). *The Demon of noontime, Ennui in Western literature*. Princeton, NJ: Princeton University Press.

Lacoue-Labarthe, P. (2003). Oedipus and figure. *Radical Philosophy*, *118*, 7–17.

Lefebvre, H. (1971). *Everyday life in the modern world*. New York: Penguin.

Lefebvre, H. (1991). *Critique of everyday life*. London: Verso.

Lefebvre, H. (1995). *Introduction to modernity*. London: Verso.

Lenoir, T. (Ed.). (1998). *Inscribing science, scientific texts and the materiality of communication*. Palo Alto, CA: Stanford University Press.

Leroi-Gourhan, A. (1993). *Gesture and speech*. Cambridge, MA: MIT Press.

Lloyd, G. E. R. (1970). *Early Greek science, Thales to Aristotle*. New York: Norton.

Lloyd, G. E. R. (2000). Observation and research. In J. Brunschwig & G. E. R Lloyd (Eds.), *The Greek Pursuit of knowledge* (pp. 67–91). Cambridge, MA: Belknap Press of Harvard University Press.

Lloyd, G. E. R. (2002). *The ambitions of curiosity, understanding the world in ancient Greece and China*. Cambridge: Cambridge University Press.

Lloyd, G. E. R., & Sivin, N. (2002). *The way and the word*. New Haven, CT: Yale University Press.

Loewenstein, G. (1994). The psychology of curiosity—a review and reinterpretation. *Psychological Bulletin*, *116*(1), 75–98.

Lowy, M. (2000). Marx's dialectic of progress: Closed or open? *Socialism and Democracy*, *14*, 1.

Luhman, N. (1995). *Social systems*. Palo Alto: Stanford University Press.

Macdonald, B. (1999). Marx and the figure of desire. *Rethinking Marxism*, *11*, 4.

Macherey, P. (1998). Philosophy as operation. In W. Montag (Ed.), *In a materialist way, selected essays* (pp. 28–42). London: Verso.

Machlup, F. (1962). *The production and distribution of knowledge in the United States*. Princeton, NJ: Princeton University Press.

Mandeville, B. (1997). *The fable of the bees*. Indianapolis, IN: Hackett.

Marcuse, H. (1960). *Reason and revolution*. Boston: Beacon Press.

Markoff, J. (2010, April). A new force behind an agency of wonder. *New York Times*, Science Section, p. 12.

Marx, K. (1962). The economic and philosophic manuscripts of 1844. In D. Struik (Ed.), New York: International Publishers.

Marx, K. (1992). Early writings. In L. Colettti (Ed.), pp. 345–358. New York: Penguin.

Maturana, H. R., & Varela, F. J. (1980). *Autopoesis and cognition: The realization of the living*. Dordrecht, Holland: D. Reidel.

McDonough, T. (Ed.). (2009). *The situationists and the city*. Brooklyn, NY: Verso.

McNeill, W. (1999). *The glance of the eye, Heidegger, Aristotle, and the ends of theory*. Albany, NY: State University Press of New York.

McNeill, D. (2005). *Gesture and thought*. Chicago: University of Chicago Press.

Merleau-Ponty, M. (1964). Eye and Mind. In J. Edie (Ed.), *The primacy of perception* (pp. 164–190). Bloomington, IN: Northwestern University Press.

Merleau-Ponty, M. (2002). *Phenomenology of perception*. New York: Routledge.

Merleau-Ponty, M. (1997). *The visible and the invisible*. Bloomington, IN: Northwestern University Press.

Merleau-Ponty, M. (2010). *Institution and passivity, course notes from the college de France (1954-1955)*. Bloomington, IN: Northwestern University Press.

Michael, J. (2000). *Anxious intellects*. Durham, NC: Duke University Press.

Mitcham, C. (1994). *Thinking through technology*. Chicago: University of Chicago Press.

Montag, W. (1999). *Bodies, masses, power. Spinoza and his contemporaries*. Brooklyn, NY: Verso.

Mulhern, F. (1994). Machiavelli and Us. *Radical Philosophy*, *101*, 39–42.

Mulvey, L. (1996). *Fetishism and curiosity*. Bloomington, IN: Indiana University Press.

Nancy, J.-L. (2010). *The truth of democracy*. New York: Fordham University Press.

Nicholls, C. (1997). *Somebody else: Arthur Rimbaud in Africa*. Chicago: University of Chicago Press.

Nietzsche, F. (1966). *Beyond good and evil*. Section 188, pp. 100–102. New York: Vintage.

Nietzsche, F. (1967). *The birth of tragedy*. New York: Vintage.

Nietzsche, F. (2001). *The pre-Platonic philosophers*. Champaign, IL: University of Illinois Press.

Nowotny, H. (2008). *Insatiable curiosity, innovation in a fragile future*. Cambridge, MA: MIT Press.

O'Hagan, A. (2010, February 25). Goodbye Moon. *Guardian*.

Ostrow, S. (1997). Preface. In J. Johnston (Ed.), *F. Kittler, Literature, media, information systems* (pp. ix–xii). Amsterdam: G+B Arts, Overseas Publishers Association.

Pearson, K. A. (1997). *Viroid life: Perspectives on Nietzsche and the transhuman condition*. New York: Routledge.

Pearson, K. A. (1999). *Germinal life, the difference and repetition in Deleuze*. New York: Routledge.

Petit, C. (2010, March 27). In Pursuit of the Briefest Beat. *Science News*.

Phillips, A. (1998). *The beast in the nursery*. New York: Vintage.

Piaget, J. (1951). *Plays, dreams and imitation in childhood*. New York: Norton.

Pickering, A. (1995). *The mangle of practice, time, agency and science*. Chicago: University of Chicago Press.

Plant, S. (1992). *The most radical gesture: The situationist international in a postmodern age* New York: Routledge.

Plato (1997a). Symposium. In J. Cooper (Ed.), *Plato: Complete works*. Indianapolis, IN: Hackett.

Plato (1997b). *Theatetus*. In J. Cooper (Ed.), *Complete works* (pp. 157–234). Indianapolis, IN: Hackett.
Plato (1997c). Apology. In J. Cooper (Ed.), *Plato: Complete works* (pp. 17–36). Indianapolis, IN: Hackett.
Plato (1997d). The Republic. In J. Cooper (Ed.), *Complete works* (pp. 1077–1107). Indianapolis, IN: Hackett.
Plutarch (1939). de Curiositae. In W. C. Helmbold (Ed.), *Moralia* (pp. 439a–523b). Harvard University Press.
Pomian, K. (1987). *Collectors and curiosities: Paris and Venice, 1500–1800*. Polity Press.
Posnock, R. (1991). *The trial of Curiosity, Henry James, William Jaes and the challenge of modernity*. New York: Oxford University Press.
Quniby, L. (1997). Technopression and the intricacies of cyborg flesh. *Constellations*, *4*(2), 229–247.
Rabinow, P. (1996). *Essays on thee anthropology of reason*. Princeton, NJ: Princeton University Press.
Rabinow, P. (1999). *French DNA*. Chicago, IL: University of Chicago Press.
Rajan, S. K. (2006). *Biocapital, the constitution of postgenomic life*. Durham, NC: Duke University Press.
Rajchman, J. (2000). *The Deleuze connections*. Cambridge, MA: MIT Press.
Raymo, C. (2000, January). *Strange beauty*. *Scientific American*, pp. 78–79.
Readings, B. (1996). *The university in Ruins*. Cambridge, MA: Harvard University Press.
Rhineberger, H.-J. (2010). *An epistemology of the concrete, twentieth century histories of life*. Durham, NC: Duke University Press.
Ricoeur, P. (2007). *Husserl*. Chicago: Northwestern University Press.
Riviere, P. (2010, January). And if you want to live forever. *Le Monde Diplomatique*, p. 14.
Roberts, J. (1999). Philosophizing the everyday: The philosophy of praxis and the fate of cultural studies. *Radical Philosophy*, *98*, 16–29.
Ronell, A. (2002). *Stupidity*. Chicago: University of Illinois Press.
Ross, K. (1996). *Fast cars, clean bodies, decolonization and the reordering of French culture*. Cambridge, MA: MIT Press.
Ross, K. (2002). *May '68 and its afterlives*. University of Chicago.
Ross, K., & Kaplan, A. (1987). Introduction, *Yale French Studies,* 73. special issue on Everyday Life.
Rossi, P. (1962). *Philosophy, technology and the arts in early modern Europe*. Princeton, NJ: Princeton University Press.
Rouse, J. (2005). *Heidegger on science and naturalism*. Retrieved November 17, 2005, from http://wesscholar.wesleyan.edu.
Sallis, J. (2000). *Force of imagination: The sense of the elemental*. Bloomington, IN: Indiana University Press.
Peirce, C. S. (1931–1958). *Collected papers*. (pp. I: 76). Cambridge, MA: Harvard University Press.
Seaford, R. (2004). *Money and the early Greek mind*. Cambridge: Cambridge University Press.
Seaford, R. (2009). World without limits; the Greek discovery that man could never be too rich. *Times Literary Supplement*, *14–15*, June 19.
Shapin, S. (2008). *The scientific life, a moral history of a late modern vocation*. Chicago: University of Chicago Press.
Shapiro, R., & Greenberg, G. (1998). Possible forms of life in environments very different from the earth. In J. Leslie (Ed.), *Modern cosmology and philosophy* (pp. 254–262). Amherst, New York: Prometheus.
Sheehan, W., & O'Meara, S. (2002). *Mars, the lure of the red planet*. Atlantic Highlands, NJ: Prometheus. 2002.
Simondon, G. (1992). The Genesis of the individual. In J. Crary & S. Kwinter (Eds.), *Incorporations* (pp. 297–319). New York: Zone Books.

Snell, B. (1953). *The discovery of the mind, Greek origins of European thought*. Cambridge, MA: Harvard University Press.
Sohn-Rethel, A. (1977). *Intellectual and manual labor, a critique of epistemology*. Atlantic Highlands, NJ: Humanities Press.
Soper, K. (1995). *What is nature?* London: Blackwell.
Spinoza, B. (1992). *Ethics*. Indianapolis, IN: Hackett.
Steigler, B. (1998). *Technics and time, 1, the fault of Epimetheus*. Palo Alto: Stanford University Press.
Steigler, B. (2009). *Acting out*. Palo Alto: Stanford University Press.
Stock, B. (1983). *The implications of literacy*. Princeton, NJ: Princeton University Press.
Stoler, A. (1995). *Race and the education of desire*. Durham, NC: Duke University Press.
Striker, G. (1996). *Essays on hellenistic epistemology and Ethics*. New York: Cambridge University Press.
Terrall, M. (2002). *The man who flattened the earth, maupertuis and the sciences of the enlightenment*. Chicago: University of Chicago Press.
Thacker, E. (2009). After life: *De Anima* and unhuman politics. *Radical Philosophy, 155*, May/June.
Thatcher, O. (1907). *The library of original sources: Early mediaeval age*. New York: Stanford University Research Extension.
Turkle, S. (1995). *Life on the screen, identity in the age of the internet*. New York: Touchstone.
Valery, P. (1972). *Leonardo, Poe, Mallarme*. Princeton, NJ: Princeton University Press.
Veblen, T. (1918). *The place of science in modern civilization*. New York: Huebsch.
Vernant, J.-P. (1982). *The origins of Greek thought*. New York: Columbia University Press.
Vienet, R. (1992). *Enrages and situationists in the occupation movement, France, May 68*. New York: Automedia.
Vlastos, G. (1991). *Socrates, ironist and moral philosopher* (p. 253). Ithaca, NY: Cornell University Press.
Voltaire, F. (1961a). *Curiosite*. In R. Naves (Ed.), *Dictionanaire philosophique* (pp. 74–75). Paris: Garnier
Voltaire, F. (1961b). L'Ame. In R. Naves (Ed.), *Dictionanaire philosophique* (pp. 12–17). Paris: Garnier.
Voltaire, F. (2001). *Micromegas*. In *Selected stories* (pp. 173–191). New York: Penguin.
Washburn, J. (2005). *University Inc. the corporate corruption of higher education*. New York: Basic Books.
Willinsky, J. (1998). *Learning to divide the world: Education at empire's end*. Chicago: University of Chicago Press.
Wolfe, C. (1998). *Critical environments: Postmodern theory and the pragmatics of the outside*. Minneapolis, MN: University of Minnesota Press.
Zacher, C. (1976). *Curiosity and pilgrimage: The literature of discovery in fourteenth century England*. Baltimore, MD: Johns Hopkins University Press.
Zizek, S. (1999). When the Party Commits Suicide. *New Left Review, 238*, 26–47.
Zizek, S. (2000). From history and class consciousness to the dialectic of enlightenment. *New German Critique, 81*, 107–123.
Zuss, M. (2000). Genomics: Technics and writing the "posthuman". *Dialectical Anthropology* 25.
Zuss, M. (2003). *Zoography. Rhizomes: 7. Cultural studies in emerging knowledges*. Retrieved September 24, 2003, from www.rhizomes.net

Index

MIX
Papier aus verantwortungsvollen Quellen
Paper from responsible sources
FSC® C105338

If you have any concerns about our products,
you can contact us on
ProductSafety@springernature.com

In case Publisher is established outside the EU,
the EU authorized representative is:
Springer Nature Customer Service Center GmbH
Europaplatz 3, 69115 Heidelberg, Germany

Printed by Libri Plureos GmbH
in Hamburg, Germany